Introductory Course on

FINANCIAL
MATHEMATICS

Introductory Course on
FINANCIAL
MATHEMATICS

M V Tretyakov

University of Nottingham, UK

Imperial College Press

ICP

Published by

Imperial College Press
57 Shelton Street
Covent Garden
London WC2H 9HE

Distributed by

World Scientific Publishing Co. Pte. Ltd.
5 Toh Tuck Link, Singapore 596224
USA office: 27 Warren Street, Suite 401-402, Hackensack, NJ 07601
UK office: 57 Shelton Street, Covent Garden, London WC2H 9HE

British Library Cataloguing-in-Publication Data
A catalogue record for this book is available from the British Library.

INTRODUCTORY COURSE ON FINANCIAL MATHEMATICS

ISBN 978-1-908977-38-0

Printed in Singapore by World Scientific Printers.

Preface

This book is based on a one-semester course, for undergraduate and postgraduate students, which was taught at the University of Leicester (UK) in 2004–2011. It was also the basis for a course for the MSc in Actuarial Science, which covers about one half of the CT8 'Financial Economics' syllabus and a part of CT1 'Financial Mathematics' syllabus of the Institute and Faculty of Actuaries (UK) professional exams.

The course is an elementary introduction to the basic ideas of Financial Mathematics and it mainly concentrates on discrete models. This course has almost no prerequisites except a basic knowledge of Probability, Real Analysis, Ordinary Differential Equations, Linear Algebra and some common sense. Elementary Probability is essentially, although briefly, revised within the course.

Financial Mathematics is an application of advanced mathematical and statistical methods to financial markets and financial management. Its aim is to quantify and hedge risks in the financial world. Having a knowledge of Financial Mathematics requires overcoming two hurdles:

1. Stochastic Analysis (which is the main mathematical tool in Financial Mathematics) and
2. Financial terminology, logic, theory and context.

It is always difficult to jump two hurdles at once. Therefore, the book starts with a low level of mathematics (a school-sized hurdle) with some financial terminology and logic thrown in. Necessary facts from Probability and Stochastics are introduced when they are required and when they can be illustrated by financial applications. The mathematical content is limited to what is actually needed to explain financial models considered in this course.

Several books and research articles were used in preparing this course. They are included in the references together with sources for further reading. In the text we usually do not indicate which books or articles were used for a particular section. The course's development was influenced most by Baxter and Rennie (1996); Shiryaev (1996); Kolb (2003); Cox and Rubinstein (1985); Filipovic (2009) and Shreve (2003).

In a short course it is not possible to touch on all aspects of the vast area of Financial Mathematics and the book mainly deals with simple but widely used financial derivatives for managing market risks. The length of the course is 30–35 lectures (50 minutes each). It consists of three parts. The first part (about eight to nine lectures long) introduces one of the main principles in Finance (and hence in Financial Mathematics) – no arbitrage pricing. It also introduces the main financial instruments such as forward and futures contracts, bonds and swaps, and options. This part is not mathematical. The second part (about 12–14 lectures long) of the course deals with pricing and hedging of European-type and American-type options in the discrete-time setting. Also, the concept of complete and incomplete markets is discussed. Mathematics-wise, elementary Probability is briefly revised and then discrete-time discrete-space stochastic processes used in this part for financial modelling are considered. The third part (about ten lectures long) starts with some basic modelling considerations including the efficient market hypothesis. The main result of this final part of the course is the famous Black–Scholes formula for pricing European options. It is derived in two ways. First, it is obtained as the limit of the discrete Black–Scholes formula from Part II via application of the central limit theorem. Secondly, it is derived by starting from a continuous-time price model (geometric Brownian motion), after the reader's knowledge of Stochastic Analysis is enhanced and, in particular, the Wiener process, Ito integral and stochastic differential equations are introduced. Some guidance for further study of this exciting and rapidly changing subject is given in the last chapter.

I would like to thank Chris Smerdon and Steve Upton, who typed the initial version of my lecture notes in 2005. I am grateful to Grigori N. Milstein and Maria Krivko for their support, discussions and advice. My special thanks are given to Yulia who drew most of the illustrations and helped with proofreading. This book would never be published without the strong encouragement of Alexander N. Gorban. I am grateful to the Imperial College Press editorial team, in particular to Tasha D'Cruz, for their help with completing this project. I also thank several generations of students in my financial mathematics classes, for their comments, corrections, enthusiasm and patience.

Nottingham, June 2013   *Michael V. Tretyakov*

Contents

Part III: Continuous-Time Stochastic Modelling and the Black–Scholes Formula   **151**

Chapter 1

Historical Remarks

We begin with a sketch of the history of the subject. Compared to other areas of mathematics, this is very young. A lot of research in Financial Mathematics is continuing to be carried out and there is a large demand from financial institutions for further developments. However, the origins of Financial Mathematics are in the past.

'Since traveling was onerous (and expensive), and eating, hunting and wenching generally did not fill the 17th century gentleman's day, two possibilities remained to occupy the empty hours, praying and gambling; many preferred the latter.' (Montroll and Shloringer, 1984)

People have always gambled and gambling is usually considered to be the origin of Probability theory. During the Age of Enlightenment critical thinking was applied to gambling outcomes which were beginning to be seen as more than simply 'God's decision'. The starting point of Probability theory is generally regarded to be an exchange of letters between Pascal (1632–1662) and Fermat (1601–1665) on some problems concerning gambling games. The first book on Probability was written by Christian Huygens (1629–1695) in 1657 under the title *About the Ratios in the Game of Dice*. The first mathematical book on Probability theory in the modern sense was *The Art of Making Conjectures* by Jacob Bernoulli (1622–1705), which was published posthumously in 1713. It contained not only probabilities in the context of gambling games but also application of Probability theory to some problems in Economics. So, even at its birth Probability theory was married to Economics and Finance.

As with all marriages, some periods are more successful than others. A major step forward in the stochastic treatment of the price of financial assets was made in the PhD thesis of a young Frenchman named Louis

Bachelier (1870–1946), *Théorie de la Spéculation* (Bachelier, 1900, see its English translation together with a facsimile of Bachelier's thesis and historical commentary in Davis and Etheridge, 2006). His thesis contained many results of the theory of stochastic processes as they stand today, which were only mathematically formalised later. He was essentially the first who theoretically studied Brownian motion (Wiener process), five years before Albert Einstein (Einstein, 1905). It is reasonable to speculate that if such a thesis had appeared 50 years later then Bachelier would probably have won the Nobel Prize in Economics but the world was a different place at the start of the 20th century. At this time the world was obsessed with Physics and science in general as much as it is now with money. Everybody knew Einstein's results on diffusion but Bachelier's results were forgotten by economists. Henri Poincaré (1854–1912), one of the greatest mathematicians of all time, was Bachelier's PhD advisor. Bachelier's thesis was praised by his mentor Poincaré but, partially due to the existing negative perception of Economics as an application of Mathematics, Bachelier was unable to join the Paris elite and spent his career in the provincial capital of Besançon near Switzerland in Eastern France (see, e.g., Courtault *et al.*, 2000; Jarrow and Protter, 2004). We can speculate that the reason that Bachelier was not recognised during his time is that what he did was not very interesting from an applicable point of view at the beginning of the 20th century; there was no real interest in pricing financial derivatives at that time.

In the world of mathematics Bachelier's work was known. For instance, in the famous paper that was pivotal for further development of Probability theory, A. N. Kolmogorov credits the discoveries made by Bachelier (Kolmogorov, 1931). In the 1940s Japanese mathematician Kiyosi Ito was influenced by Bachelier's work to create his famous Stochastic Calculus (see Footnote 8 on p. 78 of Jarrow and Protter, 2004), which we now call Ito Calculus and which is the basis of modern Stochastic Analysis and theory of stochastic differential equations (SDEs) (Ito, 1944). The Second World War was raging and this meant that it was still not the right time for Finance and Stochastics to be merged.

Bachelier's work was rediscovered by economists in the 1950s, mainly by Paul Samuelson (see Davis and Etheridge, 2006). Samuelson, Nobel Prize winner in Economics in 1970, acknowledged the impact of Bachelier's work on his research. By the mid 1960s Stochastic Analysis was a well-developed subject thanks to the work of distinguished mathematicians including Einstein, Markov, Wiener, Kolmogorov, Lévy, Doob, Ito, Gichman, Meyer and

Dynkin. Mark Davis and Alison Etheridge (2006) write:

'...when the connection was made in the 1960s between financial economics and the stochastic analysis of the day, it was found that the latter was so perfectly tuned to the needs of the former that no goal-oriented research programme could possibly have done better.'

In 1965, Samuelson (1965) introduced geometrical Brownian motion into Finance, the model which has played a central role in the development of Financial Mathematics. It was later used by Merton (1973) and Black and Scholes (1973) to derive their famous pricing formula and it is still used in financial institutions today.

A new economic situation emerged in the 1970s due to several events that had taken place since the 1960s. They led to major structural changes and to the growth of volatility in financial markets. Some of the most important of these events were: (i) transition from the policy of fixed cross rates between currencies to rates freely floating, which led to the financial crisis of 1973; (ii) the devaluation of the dollar against gold:

1971	$35 per ounce
1980	$570 per ounce
1984	$308 per ounce;

(iii) a decline in stock trade (the decline in the USA was deeper than during the 'Great Depression') and (iv) the global oil crisis provoked by OPEC (the Organization of the Petroleum Exporting Countries) (for further details see, e.g., Shiryaev, 1999). The market responded promptly to the changes in the economy and option and bond futures exchanges were opened. The first specialist exchange for trading option contracts was the Chicago Board Option Exchange (CBOE). It opened in 1973 and has been successful ever since. In 1995 the derivative market was worth $15 trillion. In 2005 it was $270 trillion, which was almost 34 times the size of the US public debt. The derivative market size reached $648 trillion by the end of 2011[1] while the global economy size was estimated at about $80 trillion[2].

With the growing volume of derivatives and volatility in financial markets, there was also a need to reconsider pricing methods as the old 'rule of thumb' and regression models became inadequate (Shiryaev, 1999). More amazing than the speed with which the market responded to the changes was

---

[1] According to the Bank for International Settlements (2012, p. 1).
[2] According to The World Factbook by CIA (2013).

the time it took for science to respond. Two landmark papers were published in 1973:

1. *The Pricing of Options and Corporate Liabilities* (Black and Scholes, 1973);
2. *The Theory of Rational Option Pricing* (Merton, 1973).

They revolutionised changed the pricing methods. Their results were immediately taken on board by option traders around the world. These papers were also, in a sense, the birth of modern Financial Mathematics which has been developed quickly and occupies a lot of researchers today. More on the history of Stochastic Analysis and its application to Financial Economics can be found in, e.g., Paul and Baschnagel (1999); Courtault *et al.* (2000); Jarrow and Protter (2004); Davis and Etheridge (2006) and Shiryaev (1999).

As we have discussed above, the merging of theory of Finance and Mathematics, in particular Stochastic Analysis, is the foundation for modern financial markets. However, there is also a third important element: computational power, without which it is impossible to quickly price financial products, evaluate risk of large portfolios, etc. Significant computational power available to financial institutions today is vital for effective financial management.

Part I

Financial Instruments and Arbitrage

This story we'll start with a riddle,
Even Alice will struggle to answer
What would remain from a fairy tale,
After it has been told?
Where, for example, is the magic horn?
Or the good fairy, where did she go?
Eh? Oh! That's the dilemma, my friend,
And in this is the whole point.[1]

In this first part of the course we will learn about one of the main principles in Finance (and hence in Financial Mathematics) – *no arbitrage pricing*. We will also become familiar with main financial instruments which include forward and futures contracts, bonds and swaps, and also options. This part is not mathematical and can be easily taught at a secondary school but it requires a lot of common sense and imagination.

---

[1] Translated from a song written by Vladimir Visotsky in 1973 for an audioplay based on Lewis Carroll's *Alice in Wonderland*.

Chapter 2

Preliminary Examples

2.1 Lesson 1 'The Expected Worth of Something is not a Good Guide to its Price'

We give two simple examples which at first glance might not seem terribly relevant to the subject; however, they are very useful (especially due to their simplicity) for understanding important concepts in Financial Mathematics introduced later in the book. In connection with these examples see also Baxter and Rennie (1996).

Example 2.1. *The Bookmaker.* A bookmaker is taking bets on a 'two-horse race', namely the FA Cup final between Liverpool and West Ham United (this example was made up before the 2006 FA Cup final). He is very clever and knowledgeable. After thoroughly studying the two teams he correctly calculates that West Ham have a 25% chance of winning while Liverpool have a 75% chance. Accordingly the odds are set at 3-1 against and 3-1 on, respectively[1]. This means that if you put a £1 stake on West Ham to win and they do triumph you get $3 + 1 = 4$ pounds; if Liverpool win you will lose £1. Betting £3 on Liverpool to win means you will get $1 + 3 = 4$ if they do win but lose £3 if they do not.

Assume that in total £15,000 is put on West Ham to win while £30,000 is put on Liverpool[2]. If Liverpool win, the bookmaker will make a net profit

---

[1] A price quoted in the form $n\text{-}m$ against (e.g., 3-1 against) means that a successful bet of £m will be rewarded with £n plus the stake returned.

[2] Of course, this is a simplification for illustration purpose only since a real bookmaker does not *a priori* know the amount of bets to be made. For further reading on the business of bookmakers, one can use, e.g., Boyle (2006).

of

$$(30000 + 15000) - (30000 + \frac{1}{3}30000) = £5000.$$

But if West Ham win then the profit is

$$45000 - (15000 + 3 \times 15000) = -£15000.$$

That is, the bookmaker makes a loss. Hence his expected profit is

$$25\%(-15000) + 75\%(5000) = £0.$$

Theoretically this seems OK. Indeed, over a number of similar and independent games, the law of large numbers (see it later in this section) would allow the bookmaker to break even. Life, however, is finite and the game cannot be repeated. The bookmaker needs money to live and run his business and until the averaging works out there is a chance of him having a substantial loss. He cannot afford to take this risk and so he, being a clever man, tries a different route in setting the odds.

His aim is now to break even no matter the result of the match. He wants:

Liverpool win $45000 - (30000 + \frac{1}{n}30000) = 0,$

West Ham win $45000 - (15000 + m \times 15000) = 0.$

The odds are now set according to the amount of money wagered. In this case 2-1 against and 2-1 on. These odds carry the *implied probability* that Liverpool have a $2/3$ chance of winning. By using these odds, the outcome of the game has become irrelevant to the bookmaker – there is *no risk* involved. In setting these new odds, the bookmaker is not looking on the probabilities of win or loss he scientifically found. While in the first case the odds related to the actual probabilities, in the second case the odds are derived from the amounts of money wagered in order to avoid any risk.

Remark 2.1. (*Not important for this course*). In reality, bookmakers will sell more than 100% of the game and will shorten the odds in order to have a profit. Suppose that the bookmaker would like to have a guaranteed profit of £3,000. To this end he will change the odds slightly as it is demonstrated in Table 2.1.

Example 2.2. *Another Game.* We play a game of coin toss where we are paid £1 for heads and nothing for tails. What price should you pay per toss to enter such a game?

Table 2.1   Selling more than 100% of the game.

	West Ham	Liverpool	
Actual Probability	25%	75%	
Bets	£15000	£30000	
Quoted Odds	9-5	5-2	
Implied Probability	$\simeq 36\%$	$\simeq 71\%$	$= 107\%!$
Profit	£3000	£3000	

Suppose that the coin is fair, i.e. heads and tails are equally likely: $P(\text{`H'}) = P(\text{`T'}) = 1/2$. Then about half the time you should win the pound and the rest of the time you get nothing. Over enough plays, you expect to make about 50p a go. So, paying more than 50p does not seem fair. Fifty pence is also the expected profit from a single toss under a formal definition of expectation. Indeed, the probabilistic model for this experiment is

$$\Omega = \{\text{`H'}, \text{`T'}\}, \qquad P(\text{`H'}) = P(\text{`T'}) = \frac{1}{2},$$

where Ω is the space of elementary outcomes from this experiment (the sample space) and $P(\cdot)$ are probabilities assigned to these elementary events from Ω (we will revise basics of Probability theory in Chapter 8). Introduce the random variable describing the result (in pounds) of one go:

$$\xi(\omega)\colon \Omega \to \mathbf{R}, \quad \text{so that} \quad \xi(\text{`H'}) = 1, \quad \xi(\text{`T'}) = 0,$$

which is a Bernoulli random variable. Its expectation is

$$E\xi = 1 \times P(\text{`H'}) + 0 \times P(\text{`T'}) = 50 \text{ (pence)}.$$

This formal expectation and the price of our game are related via the strong law of large numbers, often called Kolmogorov's strong law of large numbers.

Kolmogorov's strong law of large numbers. *Let $\xi_1, \xi_2, \ldots$ be a sequence of independent random numbers sampled from the same distribution, which has mean μ. Introduce the arithmetic average of the sequence up to the nth term:*

$$S_n = \frac{1}{n} \sum_{i=1}^{n} \xi_i.$$

Then $S_n \to \mu$ as $n \to \infty$ with probability 1, i.e. the arithmetical average of outcomes tends towards the expectation with certainty.

This result is consistent with our intuition. It tells us what the fair price is if we play for long enough. Over a long time 50p is fair because the expectation will work out. What happens in the short term though? Is 50p an enforceable price?

The game is offered again but now with a price of 45p. Instead of allowing any number of games, you are only able to play once. Further to this, the prize of £1 is raised to £10,000. The strong law of large numbers and common sense tell you that you could take advantage; 45p a game would ruin the game maker. However, we are only playing once. Hence the strong law does not apply here. Then the 'market' (if there is a market) in this game could deviate from the expected price if there are 'buyers' and 'sellers' who would agree with the suggested price of the game. This is more or less how it happens in real life. However, expectation price does seem to guide us to a starting price.

We will refer to these examples in future but now let us learn another lesson.

2.2   Lesson 2 'Time Value of Money'

In the coin game we played in the previous section, the game itself and the payment for it happen at the same time. Let us change the rules so that we continue to play in the coin game now but the payment will be made after you finish reading this book. However, £1 in the future (say, in a month) is somewhat less than £1 now and we need a rule for how to calculate the future payoff in terms of today's money. **Interest rates** serve as a formal reflection of devaluing money with time.

When you borrow money from a bank, you pay interest. **Interest** is a fee charged for borrowing assets and, most commonly, for borrowing money. It is a percentage charged on the principal amount for a period of a year. Or in reverse, interest is what you earn when you let somebody borrow your money.

We normally distinguish between *simple* and *compound* interest. **Simple Interest** is the interest on the principal amount. **Compound Interest** is paid on the original principal and on the accumulated past interest.

Let us illustrate these definitions. If you open a bank account paying interest r_m m times a year, then on having put an initial capital B_0

(*principal*) in N years you obtain the amount (the accumulated value) B_N as follows.

Simple interest:

$$B_N = B_0 + Nm\frac{r_m}{m}B_0 = B_0(1 + r_m N).$$

Compound interest. The accumulated value of a single investment of amount B_0 after $1/m$ of a year is

$$B_{\frac{1}{m}} = B_0 + \frac{r_m}{m}B_0 = B_0\left(1 + \frac{r_m}{m}\right);$$

the accumulated value of a single investment of amount B_0 after $2/m$ of a year is

$$B_{\frac{2}{m}} = B_{\frac{1}{m}} + \frac{r_m}{m}B_{\frac{1}{m}} = B_0\left(1 + \frac{r_m}{m}\right)^2,$$

the accumulated value of a single investment of amount B_0 after N years is

$$B_N = B_0\left(1 + \frac{r_m}{m}\right)^{mN}. \tag{2.1}$$

Note that in the above formulas r_m is expressed as fraction, not as a percentage. These formulas can be easily modified for the case of variable rate r.

The standard practice is to use compound interest, as you are aware from the day-to-day dealings with your own bank accounts. Once Albert Einstein was asked what he thought was the human race's greatest invention. He replied: 'Compound interest'. Compound interest is a powerful tool for building wealth and financial security. Over time, it can make your money grow dramatically. Simple interest grows only linearly with time whereas compounding causes exponential growth. There is an interesting illustration of the power of compound interest. If in 1626 the Native American tribe had invested 60 guilders, which they accepted in the form of goods for the sale of Manhattan, in a Dutch bank at 6.5% interest with annual compounding then in 2005 their investment would be worth over €700 billion (around US$820 billion), more than the value of the real estate in the whole New York City.

Example 2.3. Find the balance after three years if an amount of £100 is deposited in a bank paying:

1. 10% annual simple rate;
2. interest rate of 10% per annum (p.a.) with annual compounding;
3. interest rate of 10% p.a. with semiannual compounding;

4. interest rate of 10% p.a. with monthly compounding;
5. interest rate of 10% p.a. with daily compounding.

Answer. The corresponding accumulated values are:

1.

$$B_3 = B_0 + 3 \times r \times B_0 = \pounds 130.$$

2. Step-by-step calculation:

$$B_1 = B_0 + rB_0 = \pounds 110,$$
$$B_2 = B_1 + rB_1 = \pounds 121,$$
$$B_3 = B_2(1 + r) = \pounds 133.10.$$

3. Using the formula (2.1):

$$B_3 = B_0 \left(1 + \frac{r}{2}\right)^{2 \times 3} \approx \pounds 134.01.$$

4.

$$B_3 = B_0 \left(1 + \frac{r}{12}\right)^{12 \times 3} \approx \pounds 134.82.$$

5.

$$B_3 = B_0 \left(1 + \frac{r}{365}\right)^{365 \times 3} \approx \pounds 134.98.$$

We may also consider a **continuously compounded interest rate**, i.e. when interest is paid to your account continuously. To get the result, tend $m \to \infty$ in (2.1) and get:

$$B_N = B_0 e^{rN}. \tag{2.2}$$

This is our model for '*time value of money*' in this course, unless otherwise stated. More precisely, we assume that for any time $t < T$ (time horizon) the value now of £1 promised at time t is given by e^{-rt} for some constant $r > 0$. The rate r is then the continuously compounded interest rate for this period.

Example 2.4. Answer the same question as in Example 2.3 but for:

6. interest rate of 10% p.a. with continuous compounding.

Answer. The corresponding accumulated value is

$$B_3 = B_0 e^{3r} = 100 e^{0.3} \approx £134.99.$$

Let the continuously compounded interest be r p.a. Then after N years an initial investment B_0 will have accumulated to $B_N = B_0 e^{rN}$. Now let us find an *equivalent* compounded interest rate payable m times a year r_m, i.e. such r_m that after N years an initial investment B_0 will have accumulated to the same B_N. After elementary calculations, we get[3]

$$r_m = m \left(e^{\frac{r}{m}} - 1 \right).$$

Analogously, we find the continuously compounded interest equivalent to a compounded interest rate payable m times a year r_m :

$$r = m \ln \left(1 + \frac{r_m}{m} \right).$$

Example 2.5. A bank quotes you an interest rate of 5% p.a. with quarterly compounding. What is the equivalent rate with annual compounding?

Answer. An interest rate of 5% p.a. with quarterly compounding means that the balance after one year is

$$B_1 = B_0 \left(1 + \frac{0.05}{4} \right)^4,$$

where B_0 is a principal. The rate r_1 with annual compounding means that the accumulated value after one year is

$$B_1 = B_0 (1 + r_1).$$

Then the equivalent rate with annual compounding is

$$r_1 = \left(1 + \frac{0.05}{4} \right)^4 - 1 \doteq 0.05095 \quad (5.095\%).$$

To conclude (see also Fig. 2.1), the action of a positive interest rate is to grow an investment with time, i.e. one says that it *accumulates*. Or equivalently, the value of the investment shrinks as we look back through time to its initial (present) value, i.e. one says that it *discounts*.

Example 2.6. A business owner takes out a loan of \$450,000 on January 1 at a fixed-interest rate of 9% per year. It is repayable over 15 years with level payments due on each December 31 for the first 14 years and a final payment of \$50,000 due on December 31 of the 15th year. Calculate the annual payment amount.

Answer. This problem is for your self-study. For your verification, the final answer is \$56,032.

---

[3]Please derive it yourself.

Fig. 2.1   Accumulating and discounting in the case of continuously compounded interest.

2.3   Further Terminology

Let us introduce some further terminology from Finance.

Asset means anything of value. Assets can be **risky** or **nonrisky** (*riskless*). Here *risk* is understood as an uncertainty that can cause losses (e.g., of wealth). We may view a bank account as an (almost) riskless asset. Shares and commodities are examples of risky assets.

Definition 2.1. *A **market** is a 'place' where buyers and sellers exchange products. A **financial market** is a special type of market, where the traded product is, roughly speaking, money.*

On financial markets large sums of money are lent, borrowed and invested.

Definition 2.2. *A **portfolio** is a collection of financial assets.*

Chapter 3

Forwards, Futures and Arbitrage

In this chapter we will become familiar with the simplest derivatives, forwards and futures, as well as with the no-arbitrage principle which is the cornerstone in the theory of finance.

Financial market instruments can be divided into two categories.

a) **Underlying stock (underlier or underlying):** shares, bonds, commodities (energy like oil and gas, precious metals [gold, platinum, silver], metals [copper, nickel, tin, other], cocoa, coffee, sugar, grain and oilseed), foreign currencies.

b) **Their derivatives (derivative securities, contingent claims):** claims that promise some payment or delivery in the future contingent on another financial instrument (e.g., on underlying stock's behaviour).

Derivatives are used to reduce risk and also for speculation. The most widely used are forwards, futures, options and swaps. Pricing and hedging[1] derivatives is among the main problems considered in Financial Mathematics.

We note that derivatives can be written on other derivatives (one can call them 'second' derivatives) and also on non-directly financial instruments (for instance, on weather, see, e.g., Jewson and Brix, 2005).

3.1 Simple Stock Model

In our forthcoming considerations, it will be useful to have a model for stock price in mind. The widely accepted (although simplistic) model is that stock prices at a fixed time T are *log-normally distributed*. Let ξ be a normally distributed random variable with mean μ and variance σ^2: $\xi \sim \mathcal{N}(\mu, \sigma^2)$.

---

[1]We will define hedging in the very near future.

We assume that the log of the stock price S_T during a time period T changes by ξ:

$$\ln S_T = \ln S_0 + \xi \tag{3.1}$$

or

$$S_T = S_0 \exp(\xi).$$

Recall that this ξ is a continuous random variable with the probability density

$$p(x) = \frac{1}{\sqrt{2\pi\sigma^2}} e^{\frac{-(x-\mu)^2}{2\sigma^2}}.$$

This model is simple and various corrections to it exist. We note that this model satisfies the natural requirement for prices to be positive. In this course we do not consider modelling of negative prices which appear, e.g., in energy markets.

3.2   Forward Contract

We start our consideration of financial derivatives with the oldest, most natural and easiest – a forward contract.

Definition 3.1. Forward contract *(or simply **forward**) is a contract between two parties whereby one party promises to deliver to the other the stock at some agreed time T in the future (at the contract's expiry date) in exchange for an amount K agreed upon now.*

Its main (original) purpose is to share risk.

Example 3.1. Imagine a farmer who grows corn. He needs to plan his work and expenses during winter and spring before he will get corn. He needs a loan now and would like to be sure that he will be able to return it in autumn when he will sell the corn. However, the price of corn is volatile and there is a risk that the price will go down and the farmer will not be able to return the loan. As a result, his business would collapse. Obviously, he does not want to take this risk. So he goes to his friend, the miller, who offers him the following

deal. The farmer will sell to the miller a certain amount of his harvest at a price prescribed now (independent of what happens to the price of the corn in autumn). Then they can both plan their economic situation and risk is (almost) eliminated.

We can see from the above example that forwards are quite simple contracts in practice and they are beneficial for protecting business from uncertainty in the future. Forward contracts have a long history. There is evidence that Roman emperors entered forward contracts on Egyptian grain and there are traces of the use of forwards in classical Greek times and ancient India.

The aim of Financial Mathematics on this occasion is to find the price K that is fair (acceptable) for both seller and buyer (farmer and miller in our example). More precisely, the *pricing question* here is: *What amount K should be written into the forward now to pay for the stock at the contract's expiry T?*

The stock price at time t is S_t and the forward payment in the contract is K. Thus the value of the contract from the perspective of the buyer of the stock at its expiry T is $S_T - K$. In other words, the *payoff function* here has the form $f(s) = s - K$. This function is plotted in a *payoff diagram*, Fig. 3.1.

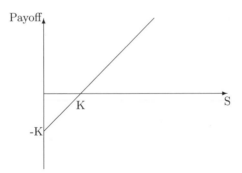

Fig. 3.1   Payoff diagram for a forward.

Since there is no payment (premium) to enter into a forward contract, we have the following *profit diagrams*. The diagram of Fig. 3.2 is for the seller of the stock (the holder of the *short forward* contract) – the farmer in Example 3.1.

The profit diagram of Fig. 3.3 is for the buyer of the stock (the holder of the *long forward* contract) – the miller in Example 3.1.

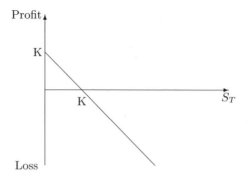

Fig. 3.2   Profit/loss diagram for a short forward contract.

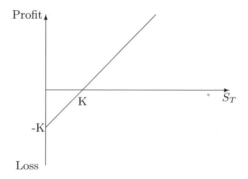

Fig. 3.3   Profit/loss diagram for a long forward contract.

Due to the 'time value of money', the value of the payoff of the long forward (miller's position) as of now is

$$e^{-rT}(S_T - K).$$

To price this derivative, let us apply common sense from both sides: the farmer and miller from Example 3.1. The strong law of large numbers from Section 2.1 suggests that the expected value of this discounted payoff should be equal to zero:

$$E[e^{-rT}(S_T - K)] = 0.$$

Indeed, if it is positive then the long-term use of this pricing mechanism leads to the miller's profit in Example 3.1. On the other hand, if it is negative then the farmer profits. They should both be happy if [2]

$$K = ES_T = ES_0 e^{\xi} = S_0 E e^{\xi} = S_0 e^{\mu + \frac{\sigma^2}{2}}$$

(we assume here that the price follows the model (3.1)). This looks very natural and corresponds to our normal everyday thinking, but is it right to use? In fact, this price could only be the market price by a pure coincidence. The answer is actually wrong and the logic based on the strong law is wrong. No farmer or miller would price a forward like that. To price the forward, we should apply a completely different logic.

Farmers and millers have relatively small businesses. As such, the markets for them can be assumed to have infinite capacity which implies that the stock can be bought and sold in unlimited amounts for the existing price and money can be borrowed and lent at the continuously compounded interest rate r in arbitrary size. Further, we assume that there is no charge for holding arbitrarily positive and negative amounts of stock and there are no transaction costs and taxes. Here 'positive' means you own the stock (in the *long position*) and 'negative' means that you are borrowing it (in the *short position*). Being in the short position (also known as short selling) can and often does happen in real life with regards to both stock and money: investors can borrow stock as well as money.

Consider the seller (the farmer in Example 3.1) of the contract, obliged to deliver the stock (grain) at time T in exchange for some agreed amount (from the miller). Instead of growing the grain himself, he could cheat. He can go to a bank and borrow S_0 now and buy the stock with it. He can then store the stock over summer and relax instead of working the fields. When the contract expires, he has to pay the loan back to the value of $S_0 e^{rT}$ and has the stock ready to deliver. If the forward price K was defined such that $K < S_0 e^{rT}$ then the seller (the farmer) will make a loss with certainty. He is unlikely to let this happen so will demand that the forward price is $K \geq S_0 e^{rT}$.

Meanwhile, the buyer (the miller) can perform a scheme similar to the farmer's. He can go to a bank to borrow S_0 and buy the stock ready for the autumn. At the end of the contract, the stock will have cost him $S_0 e^{rT}$. Therefore, he will not be prepared to pay any more than this if buying stock from the seller (the farmer). Hence the price from the buyer's point of view should be $K \leq S_0 e^{rT}$.

---

[2] If you are not familiar with the formula $Ee^{\xi} = e^{\mu + \frac{\sigma^2}{2}}$ for a Gaussian random variable ξ with mean μ and variance σ^2, then you are encouraged to do this simple exercise and prove this relation yourself.

Combining these two points of view we get:

$$S_0 e^{rT} \leq K \leq S_0 e^{rT}$$

and, consequently, the *forward price* should be equal to

$$K = S_0 e^{rT}. \tag{3.2}$$

This price does not depend on the expected value of the stock, it does not even depend on the stock price having a particular distribution (remember Example 2.1 about the bookmaker?). Any attempt to offer a different price on a market would allow someone to take advantage via the construction of a procedure such as that explained above. So, why does the strong law fail here? As we considered in the coin game (see Example 2.2), the strong law cannot enforce a price in the short run, it only suggests. In the case of forwards, a completely different mechanism enforces the price. Through this simple financial instrument, we come to the very important notion of the theory of finance – *arbitrage*.

Remark 3.1. (*Important*). Note that in finding the forward price (3.2) we used **two** instruments: grain and money which allowed us to discover the arbitrage price (cf. the second part of Example 2.2 where we had just one instrument [money] and could not enforce a price, i.e. could not find an arbitrage price).

3.3   Arbitrage

Definition 3.2. *Arbitrage means making of a guaranteed risk-free profit with a trade or series of trades in the market.*

Example 3.2. In Example 3.1 an attempt to use a price not equal to $S_0 e^{rT}$ is an arbitrage since this leads to somebody taking advantage and making (unlimited) riskless profit.

Definition 3.3. *An* **arbitrage-free** *(no arbitrage opportunity) market is a market which has no opportunities for risk-free profit.*

Definition 3.4. *An* **arbitrage price** *is a price for a security that allows no arbitrage opportunity (i.e. does not allow guaranteed risk-free profit).*

Example 3.3. The price $K = S_0 e^{rT}$ is an arbitrage price for forwards.

The notion of arbitrage plays a crucial role in Finance; it is fundamental for everything we do in Financial Mathematics. In what follows, our aim will always be to find a fair price of financial instruments under the arbitrage-free condition. In a sense arbitrage is the central notion for this whole course.

Let me confuse you a little bit more. Given the definitions of arbitrage and an arbitrage price, we can now say that the strong law of large numbers was not wrong when we used it to price a forward earlier. If we believe in the price model (3.1) and, for example, $S_0 e^{\mu + \frac{\sigma^2}{2}} > S_0 e^{rT}$ then a buyer of stock in a forward contract (he has a long forward contract) *expects* to make money. However, an arbitrage price overrides the expected price found via application of the strong law of large numbers. In other words, if there is an arbitrage price then any other price is unfair towards one of the parties. At the same time, we also note that the expected price can motivate the buyer from the above example to speculate.

Example 3.4. A one-year *long forward contract* (this party agrees to buy) on a commodity is entered into when the commodity price is £40 and the fixed-interest rate is 10% p.a. with continuous compounding.

1. What is the forward price?
2. What is the initial value of the forward contract?
3. Six months later, the price of the commodity is £45 (the interest rate stays the same). What is the present forward price and the present value of the original contract?

Answer.

1. Forward price is $K = S_0 e^{rT} = 40e^{0.1} \approx £44.21$.
2. Initial price is zero, we are not paying any premium to enter into this contract.
3. The present forward price is $K_2 = S_{\frac{1}{2}} e^{r(T-t)} = 45e^{0.05} \approx £47.31$. The value of the forward contract is the amount you would have to pay the holder to buy it from him. According to the original long forward, the value of the claim at maturity will be

$$S_T - 40e^{0.1}.$$

While for the new (current) forward it is

$$S_T - 45e^{0.05}.$$

The difference is $45e^{0.05} - 40e^{0.1}$ and, applying the 'time value of money', the fair present value (price) of the original forward is

$$e^{-0.05}(45e^{0.05} - 40e^{0.1}) = 45 - 40e^{0.05} \doteq £2.95.$$

Here we used the fact that the fair present price of the new forward is zero since we are paying no premium to enter into a forward.

Remark 3.2. When we considered pricing forwards above (in Examples 3.1–3.4), we made a simplification and omitted the further three factors which affect the forward price in reality. First, we assumed that the asset did not provide a cash income or yield to its holder (e.g., a stock paying dividend provides such an income). Second, we neglected possible storage costs arising in the case of commodities. Finally, in the case of consumption commodities (like oil, grain, copper, cocoa, etc.) one has to take into account convenience yields[3] when pricing forwards. It is not difficult to generalise the forward pricing formula by incorporating these three additional possible factors (see, e.g., Kolb, 2003; Hull, 2003). We just need to always keep in mind the no-arbitrage principle.

Example 3.5. Suppose one enters into a forward contract with maturity in T years on a commodity when its price is S_0 per unit and the fixed-interest rate is r p.a. with continuous compounding. Further, let C be the present value of all the storage costs per unit of this commodity that will be incurred during the life of the forward contract. Using the no arbitrage arguments, find the corresponding forward price.

Answer. This problem is for your self-study. For your verification: after using arbitrage-free arguments, you should eventually obtain that the forward price is equal to $K = (S_0 + C)e^{rT}$.

3.4   Futures Contract (Futures)

Definition 3.5. *A **futures** is a forward traded in a formalised exchange.*

Forwards are custom made and are traded *over the counter* (OTC) between two counterparties. A futures contract is a forward contract in which every aspect is standardised.

---

[3]A convenience yield is defined as the amount of benefit that is associated with physically holding a particular commodity rather than having a forward or option for it (e.g., one usually does not consider a forward contract on crude oil to be equivalent to crude oil held in inventory). The physical crude oil, in contrast to a forward contract on it, can be an input to, e.g., a refining process. In the case of consumption commodities, ownership of a physical asset enables a manufacturer to keep their production running and also possibly profit from temporary shortages.

The main differences between forwards and futures are:

1. Futures contracts always trade on an organised exchange (e.g., the Chicago Board of Trade since 1848).
2. Futures are always highly standardised with a specified quantity of a good, a specified delivery date and delivery mechanism.
3. Performance on futures is guaranteed by a *clearing house* (a financial institution associated with the futures exchange that guarantees the financial integrity of the market to all traders).
4. All futures require that traders post margins in order to trade. A *margin* is a deposit made by the futures trader to guarantee his financial obligations that may arise from the trade.
5. Futures markets are regulated by a government agency, while forward contracts generally trade in an unregulated market.

Forwards and futures are similar contracts which differ in the way they are traded. Their pricing and use are similar (see further details in Kolb, 2003; Hull, 2003).

In the next definition we introduce a new notion, hedging, though we have illustrated it already in Example 3.1.

Definition 3.6. *To* **hedge** *means to protect a position against the risk of market movements.*

Futures are used for the following three purposes:

1. Hedging (to reduce risks).
2. To obtain information about the future prices of an asset.
3. Speculation.

Hedging using futures works in the same way as in the case of forwards which we illustrated in Example 3.1. The second use of futures is illustrated in Example 3.6 below. The first two uses are usually considered as socially useful whilst the last one is often not viewed as socially useful. However, speculators are important market participants because they add liquidity to the market without which the other, 'social' uses cannot function (see further examples, e.g., in Kolb, 2003; Hull, 2003).

Example 3.6. An investor is looking at the possibility reopening a coal mine. The decision to start up the mine relies on the price the miner will receive for coal. If the investor puts her money into the mine business now, the production would only start in 12 months time. How would she predict that this project will be financially viable? The investor can look at the price quoted now in the futures market for a contract on coal with delivery in 12 months. If the price is high enough, she can justify reopening the mine. In this situation the investor has used the futures market for price discovery.

Futures provide us with estimates of future prices of assets that are usually considered as one of the best forecasts.

Remark 3.3. Note that we have considered the pricing of forward and futures contracts in perfect markets. In real markets there are some imperfections that affect pricing (Kolb, 2003).

Chapter 4

Bonds and Swaps

On the financial market there are instruments (securities) which play the role of an (almost) riskless asset – **bonds**. Bonds are interest-bearing securities, or in other words, bonds are promissory notes issued by a government, bank or other financial establishment to raise capital. The interest on bonds is payable on a regular basis and the repayment of the entire loan (i.e. the *principal* or the *face value* of the bond) at a specified time (*exercise* or *redemption* or *maturity time*) is guaranteed. Instead of putting money in a bank, you can buy a piece of paper called a bond and earn interest in a similar way as on your bank account. In reality, bonds are not risk free and their price depends not only on the time value of money but also on the credibility of the promiser. Government bonds usually have better protection and they are usually less risky than corporate ones (Kolb, 2003). However, in this course we will only concern ourselves with the time value of money for default-free borrowing. For further reading on interest rate modelling including modelling of defaultable bonds, see, e.g., Brigo and Mercurio (2006); Björk (2004) and Filipovic (2009).

Governments and corporations use bonds to raise capital. The two other possible ways of raising capital by corporations are to issue shares or to take a loan from a bank (or another financial institution). The choice of using one or another instrument or their combination depends on the financial circumstances.

4.1 Zero-Coupon Bonds and Interest Rates

The basic notion used in interest-rate modelling is *default-free zero-coupon discount bond* (also called *discount bond*), which is an agreement to pay some money $P(0, T)$ now (time $t = 0$) with the promise of receiving \$1 (or one unit of another currency) at the maturity date T. It is called a zero-coupon (in contrast to coupon bonds considered in the next section) because the only cash exchange that takes place is at the end of the life of this fixed-income instrument, i.e. at the maturity date T, see Fig. 4.1.

In theoretical considerations one assumes that (i) there is a frictionless market[1] for bonds with any maturity $T > 0$; (ii) $P(T,T) = 1$ and (iii) the function $P(t,T)$ is differentiable in the maturity time T.

On real financial markets bonds have maturities only at specific dates, i.e. they do not exist for all maturities T and our assumption (i) is not valid in practice. Having bond prices $P(t,T_i)$ for a set of maturities T_0, $T_1, \ldots$, T_n, we can reconstruct a function $P(t,T)$ using an interpolation, which is usually done for purposes of financial analysis and modelling. In addition, real fixed-income markets are not frictionless, e.g., there are transaction costs.

The assumption (ii) does not take into account the possibility of default of the bond issuer. In the case of default the bond owner can recover only part of the bond value, i.e. $P(T,T)$ becomes less than 1 (possibly 0). However, we will use these assumptions here since they are appropriate for our goal of learning some of the basics of interest-rate markets.

$P(t,T)$ $\quad\quad\quad\quad\quad\quad\quad\quad\quad\quad\quad\quad\quad\quad\quad\quad$ $P(T,T) = 1$

t $\quad\quad\quad\quad\quad\quad\quad\quad\quad\quad\quad\quad\quad\quad\quad\quad\quad\quad\quad\quad\quad$ T

Fig. 4.1   Zero-coupon bond.

The third condition is technical, it ensures that the *term structure of default-free zero-coupon discount bond prices* (which is also called *discount curve*) $T \longmapsto P(t,T)$ is a smooth curve. A typical dependence of $P(t,T)$ on T for a fixed t is given in Fig. 4.2.

It is clear that before time t we do not know the term structure $P(t,T)$ with certainty and hence $t \longmapsto P(t,T)$ is a stochastic process (we will rigor-

---

[1]Frictionless market is a theoretical trading environment where there are no costs and restraints associated with transactions. In particular, in the case of bonds it means that the price for a bond does not change when its owner changes.

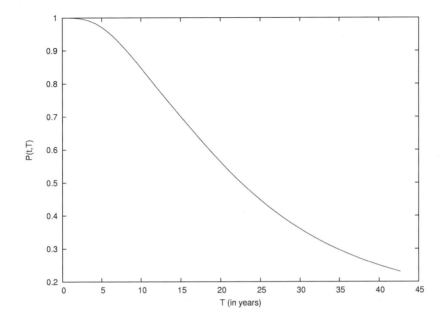

Fig. 4.2   Term structure of UK gilts on 20th July 2012.

ously introduce the notion of stochastic processes and the associated information flow in Chapter 9). An illustration of a typical trajectory of $P(t, T)$ for a fixed T is given in Fig. 4.3.

To understand the term structure behaviour, one usually uses *implied interest rates*, which better visualise the information about term structure than the function of two variables $P(t, T)$. There is a whole 'zoo' of interest rates which we will now introduce.

To this end, we first consider a *forward rate agreement* (FRA). This is a contract with a specified fixed-interest rate which will apply in borrowing or lending a notional cash sum for an agreed period in the future. The contract involves three dates: the current time s, the start of the loan $t \geq s$ and the maturity date of the loan $T > t$.

Example 4.1. Knowing only today's term structure $P(s, \cdot)$, what at time T is the value of investing \$1 at a date $t \in [s, T]$? To discover this value, we use the following strategy:

- At time s (i.e. today) sell one zero-coupon bond with maturity $t \geq s$ for \$$P(s, t)$ and, using the generated sum of money $P(s, t)$, buy

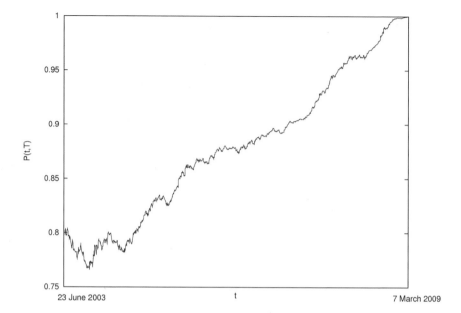

Fig. 4.3   Calender time dependence of the 4% UK Treasury Principal Strip between 23 June 2003 and 7 March 2009.

$P(s,t)/P(s,T)$ of zero-coupon bonds with maturity $T \geq t$; the value of this portfolio is zero.

- At time t, pay \$1 according to the bond with maturity t we sold at time s, i.e. we have *invested* \$1 at time t.
- At time T, receive $\$P(s,t)/P(s,T)$ according to the bonds with maturity T we bought at time s.

Thus, *as for today – time* s, the value of the forward investment of \$1 made at time t becomes $\$P(s,t)/P(s,T)$ at time T with certainty or, in other words, the future investment of $\$P(s,T)/P(s,t)$ made at time t yields \$1 at time T. Note that any value other than $P(s,T)/P(s,t)$ will lead to arbitrage (we will be able to extract riskless profit).

Having the above example in mind, we introduce the following notions:

I1 The *simple* (or simply compounding) *forward rate* at time $s \leq t$ for the period $[t,T]$ is given by

$$F(s;t,T) = \frac{1}{T-t}\left(\frac{P(s,t)}{P(s,T)} - 1\right), \qquad (4.1)$$

which is equivalent to

$$\frac{P(s,t)}{P(s,T)} = 1 + (T-t)F(s;t,T) \quad \text{or} \quad 1 = \frac{P(s,T)}{P(s,t)}\left[1 + (T-t)F(s;t,T)\right],$$

i.e. $F(s;t,T)$ is, as for the time moment s, the simply compounding interest rate (see Section 2.2) for the investment of $\$P(s,T)/P(s,t)$ to be made at time t in the future to yield $\$1$ at time T.

I2 The *simple spot rate* for the period $[t,T]$ is given by

$$F(t;T) = F(t;t,T) = \frac{1}{T-t}\left(\frac{1}{P(t,T)} - 1\right), \tag{4.2}$$

i.e. $F(t;T)$ is, as for the time moment t, the simply compounding interest rate for the investment of $\$P(t,T)$ made at time t to yield $\$1$ at time T.

I3 The *continuously compounded forward rate* at time $s \le t$ for the period $[t,T]$ is given by

$$R(s;t,T) = -\frac{\ln P(s,T) - \ln P(s,t)}{T-t}, \tag{4.3}$$

which is equivalent to

$$\frac{P(s,t)}{P(s,T)} = \exp\left((T-t)R(s;t,T)\right) \quad \text{or} \quad 1 = \frac{P(s,T)}{P(s,t)}\exp\left((T-t)R(s;t,T)\right),$$

i.e. $R(s;t,T)$ is, as for the time moment s, the continuously compounding interest rate (cf. (2.2)) for the investment of $\$P(s,T)/P(s,t)$ to be made at time t in future which yields $\$1$ at time T.

I4 The *continuously compounded spot rate* for the period $[t,T]$ is given by

$$R(t;T) = R(t;t,T) = -\frac{\ln P(t,T)}{T-t}, \tag{4.4}$$

i.e. $R(t;T)$ is, as for the time moment t, the continuously compounding (fixed) interest rate at which an investment of $\$P(t,T)$ at time t accumulates continuously to yield $\$1$ at time T. It is sometimes called *zero rate*.

The continuously compounded spot rate $R(t;T)$ gives *zero-coupon yield*, i.e.

$$P(t,T) = \exp(-(T-t)R(t;T)).$$

Accordingly, the function $T \longmapsto R(t;T)$ is called the *(zero-coupon) yield curve*[2].

---

[2]Note that the term 'yield curve' is ambiguous and used differently in different textbooks.

I5 The *instantaneous forward rate* at time s with maturity $t \geq s$ is defined by

$$f(s,t) = \lim_{T \to t} R(s;t,T) = -\frac{\partial}{\partial t} \ln P(s,t). \tag{4.5}$$

Note that $f(s,t)$, $s \leq t$, represents the instantaneous continuously compounded rate prevailing at time s for riskless borrowing or lending over the infinitesimal time interval $[t, t+dt]$ in future.

Using the instantaneous forward rate and recalling that $P(T,T) = 1$, we can express the bond price as

$$P(t,T) = \exp\left(-\int_t^T f(t,u)du\right). \tag{4.6}$$

Exercise 4.1. Explain why $\lim_{T \to t} R(s;t,T) = -\frac{\partial}{\partial t} \ln P(s,t)$.

Exercise 4.2. Explain why (4.6) follows from (4.5).

For a fixed t, the instantaneous forward rate $f(s,t)$ is a function of the maturity t, which is usually called the *forward curve* at time s.

I6 The *instantaneous spot rate* (or short rate) at time s is defined by

$$r(s) = f(s,s) = \lim_{t \to s} R(s,t).$$

We can say that the instantaneous spot rate is the risk-free rate of return at time s over the infinitesimal interval $[s, s+ds]$. The instantaneous spot rate is just a process in time, free of any other parameters and it does not specify the entire yield curve.

The interest rates we introduced in I1–I6 are called *implied interest rates*. They are *implied* by the term structure $P(t,\cdot)$, observable on the market at time t.

The return of \$1 at time s over the time interval $[s, s+\Delta s]$ is equal to[3]

$$\frac{1}{P(s,s+\Delta s)} = \exp\left(\int_s^{s+\Delta s} f(s,u)du\right) = 1 + r(s)\Delta s + o(\Delta s). \tag{4.7}$$

Instantaneous re-investment of the money $1/P(s,s+\Delta s)$ at time $s+\Delta s$ in bonds maturing at $s + 2\Delta s$ gives at time $s + 2\Delta s$:

$$\frac{1}{P(s,s+\Delta s)}\frac{1}{P(s+\Delta s,s+2\Delta s)} = (1 + r(s)\Delta s)(1 + r(s+\Delta s)\Delta s) + o(\Delta s),$$

---

[3] As usual, $o(\Delta s)$ means that $o(\Delta s)/\Delta s \to 0$ as $\Delta s \to 0$.

and so on, which can be viewed as the strategy of 'rolling over' of the maturing bonds. The limit[4] of this strategy as $\Delta s \to 0$ is called *bank account* (*money-market account, savings account* or *money account*) $B(s)$. Then, according to (4.7), it instantaneously grows at time s at the short rate $r(s)$:

$$B(s + \Delta s) = B(s) + r(s)B(s)\Delta s + o(\Delta s),$$

which in the limit of $\Delta s \to 0$ becomes

$$dB = r(s)B(s)ds, \ B(0) = B_0. \tag{4.8}$$

Solving the above ordinary differential equation, we get

$$B(t) = B_0 \exp\left(\int_0^t r(s)ds\right). \tag{4.9}$$

The interpretation of the money-market account is a strategy of instantaneously re-investing money at the current short rate (think about how your savings bank account works).

Note that we get (2.2) from (4.9) by assuming constant short rate

$$r(s) = r \tag{4.10}$$

for the period of time under consideration.

The bank account $B(t)$ defined in (4.8) (or (4.9)) relates amounts of money available at different times, i.e. it allows us to answer the question: what was the value at time t of \$1 (or in general, one unit of a currency) available at time T? Indeed, we have \$1 in the money-market account at time T if at time $t \le T$ we had

$$D(t,T) = \frac{B(t)}{B(T)} = \exp\left(-\int_t^T r(s)ds\right)$$

dollars in the money-market account. Let us emphasise that in real life the discount factor $D(t,T)$ is **not** known with certainty at time t (because $r(s)$ is known at time s and not known before s) in comparison with the discounting factor $P(t,T)$ (i.e. the bond price) which is always known with certainty at time t. Note that both $P(t,T)$ and $D(t,T)$ give the value at time t of \$1 available at time T; the fundamental difference is that $P(t,T)$ is the 'fair' price given information about the market until the time t while $D(t,T)$ is the price based on the information until the time T[5].

Example 4.2. If short rate is assumed to be deterministic (in particular, if the simplified assumption (4.10) holds), then we have $D(t,T) = P(t,T)$.

---

[4]See details in Björk *et al.* (1997).

[5]For a precise relation between $D(t,T)$ and $P(t,T)$ see, e.g., Brigo and Mercurio (2006) and Filipovic (2009), where it is shown that $P(t,T)$ is equal to expectation of $D(t,T)$ conditioned on the information available at time t taken with respect to the risk-neutral probability measure. We do not consider this question in this course.

Before we justify the claim made in the above example, let us consider another one.

Example 4.3. In the deterministic world[6] (i.e. when we know at time s_0 bond prices $P(s,t)$ with certainty not only for earlier times $s \leq s_0$ but also for all future times $s_0 < s \leq t$) the no-arbitrage principle implies that

$$P(s,T) = P(s,t)P(t,T), \quad s \leq t \leq T. \tag{4.11}$$

Answer. We prove the equality (4.11) by contradiction. First assume that $P(s,T) > P(s,t)P(t,T)$ and find a strategy which will make us a riskless profit. Indeed, let us consider the following strategy:

- At time s sell $P(s,t)P(t,T)/P(s,T)$ of today's bonds with maturity at time T and buy $P(t,T)$ of today's bonds with maturity at time t (note that we know $P(t,T)$ now because of our assumption that the world is deterministic). This initial portfolio has the value $V_s = 0$.
- At time t, receive $\$P(t,T)$ according to the bought bonds with the maturity t in the previous step and using them buy one of today's bonds with maturity T. Note that we neither invest additional funds nor consume funds at this time, we just re-invest.
- At time T, we receive $\$1$ (thanks to the bought bond in the previous step) and should pay $\$P(s,t)P(t,T)/P(s,T)$ according to the sold bonds of maturity T in the first step, i.e. our portfolio has the value $V_T = 1 - P(s,t)P(t,T)/P(s,T)$ which is strictly positive due to the assumption made. Hence, we made a riskless profit which contradicts the no-arbitrage condition and, consequently, the assumption made is false.

To complete the proof of (4.11), we need to analogously consider the case $P(s,T) < P(s,t)P(t,T)$.

Exercise 4.3. Complete the proof in Example 4.3.

Answer for Example 4.2. We take logarithms of both sides in (4.11), recall (4.6), do simple manipulations with the integrals and arrive at (please check):

$$\int_t^T f(s,u)du = \int_t^T f(t,u)du, \quad s \leq t \leq T.$$

---

[6]It goes without saying that this assumption is purely theoretical.

Then by the fundamental theorem of calculus[7], $f(s,T) = f(t,T)$ for all $s \le t \le T$ and hence $f(s,T) = f(T,T) = r(T)$ for $s \le T$. Thus

$$P(t,T) = \exp\left(-\int_t^T f(t,u)du\right) = \exp\left(-\int_t^T r(u)du\right) = D(t,T), \quad t \le T.$$

Remark 4.1. In this course (except in this chapter), we will only consider non-random interest rates; usually that the simplified assumption (4.10) is valid (unless otherwise explicitly stated). Hence, we will not distinguish between bonds and bank accounts. This is a simplification which will be used in the following chapters of the course to explain the basic ideas of financial products on equity markets in a clear way. This simplified assumption is often used in conjunction with the understanding that variability of interest rates contributes much less to prices of equity derivatives than their underlier's movements. For further reading on interest rate modelling see, e.g., the textbooks by Baxter and Rennie (1996) and Filipovic (2009) and also more comprehensive books by Andersen and Piterbarg (2010) and Brigo and Mercurio (2006).

Example 4.4. Suppose continuously compounded spot rates are as in Table 4.1.

Table 4.1 The continuously compounded spot rates used in Example 4.4.

Maturity in years	Rate, % p.a.
1	6
2	5.5
3	5.2
4	5.0
5	4.8

Find the continuously compounded forward rates for the second, third, fourth and fifth years.

Answer. Due to the relations (4.3) and (4.4), we have (please check):

$$R(0; t, T) = \frac{T}{T-t}R(0; T) - \frac{t}{T-t}R(0; t). \tag{4.12}$$

Hence, the continuously compounded forward rates for the second, third, fourth and fifth years are $R(0; 1, 2) = 5.0\%$, $R(0; 2, 3) = 4.6\%$, $R(0; 3, 4) = 4.4\%$ and $R(0; 4, 5) = 4.0\%$, respectively.

[7]It is assumed that $f(t,u)$ is continuous in u.

4.2   Coupon Bonds

On fixed-income markets, the amount of zero-coupon bonds is relatively small and bonds usually have coupons. We distinguish coupon bonds, in which periodic payments are fixed at the time of their issue (fixed coupon bonds), from those in which payments vary (floating rate notes) according to some benchmark interest rate. We start with the simpler product of the two.

Definition 4.1. *A (fixed) **coupon bond** is a contract specified by its nominal (or face) value N_0, a number of dates $T_1 < \cdots < T_n$ (the coupon dates) and a sequence of (deterministic) coupons $c_1, \ldots, c_n$ so that the owner of the bond receives c_i from its issuer at time T_i, $i = 1, \ldots, n$, and the nominal value N_0 at the maturity (terminal) time $T = T_n$.*

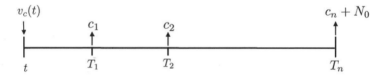

Fig. 4.4   Cash flow for a coupon bond.

The cash flow for a coupon bond is illustrated in Fig. 4.4 and the initial price $v_c(t)$, $t \le T_1$, of a coupon bond is equal to the sum of discounted cash flows:

$$v_c(t) = \sum_{i=1}^{n} c_i P(t, T_i) + P(t, T_n) N_0. \tag{4.13}$$

Exercise 4.4. Show that (4.13) is true.

Usually, the coupon dates are equally spaced: $T_{i+1} - T_i = \delta$ for $i = 1, \ldots, n - 1$, and the coupons are quoted as a fixed percentage of the face value: $c_i = \rho \delta N_0$ for a fixed-interest rate ρ. Then the formula (4.13) becomes

$$v_c(t) = \left(\rho \delta \sum_{i=1}^{n} P(t, T_i) + P(t, T_n) \right) N_0. \tag{4.14}$$

Note that if $\rho = 0$ the coupon bond reduces to a zero-coupon bond with maturity T_n.

Example 4.5. Find the price of a coupon bond with face value \$100 and \$2 semiannual coupons (the first coupon is to be paid in six months) that matures in three years given the current instantaneous forward rate curve $f(0, T) = \ln(150 + 48T)/100$.

Answer. Due to (4.13) and (4.6), we have

$$v_c(0) = 2 \sum_{i=1}^{6} P(0, 0.5i) + 100 P(0, 3)$$

$$= 2 \sum_{i=1}^{6} \exp\left(-0.01 \int_0^{0.5i} \ln(150 + 48u) du\right)$$

$$+ 100 \exp\left(-0.01 \int_0^3 \ln(150 + 48u) du\right)$$

$$\approx 96.03 \ (\$).$$

Now let us consider bonds which pay coupons that are reset for every coupon period. Usually, the resetting is linked to a benchmark interest rate. One of the main examples of such interest rates is the *London Interbank Offer Rate*[8] (LIBOR for short). This is a benchmark rate derived from interbank lending rates reported to the British Banker's Association (BBA) by a number of banks and it is used as a reference rate for many fixed-income products. It has a series of possible maturities (from overnight to 12 months) and is quoted on a simple compounding basis. Let us define bonds with variable coupons called floating rate notes.

Definition 4.2. A *floating rate note* is specified by its nominal (or face) value N_0, a number of dates $T_0 < T_1 < \cdots < T_n$ (the coupon dates), and a sequence of coupons $\varsigma_1, \ldots, \varsigma_n$ such that

$$\varsigma_i = (T_i - T_{i-1}) F(T_{i-1}, T_i) N_0, \quad i = 1, \ldots, n, \tag{4.15}$$

where $F(T_{i-1}, T_i)$ is the simple spot rate. The owner of the floating rate note receives ς_i from its issuer at time T_i, $i = 1, \ldots, n$, and the nominal value N_0 at the maturity (terminal) time $T = T_n$.

---

[8]E.g., see BBA's LIBOR website at http://www.bbalibor.com/ and further details in Brigo and Mercurio (2006); Hull (2003); Schoenmakers (2005). We also note that following the 2012–2013 LIBOR scandal it is expected that there will be a major change in how LIBOR rates are set, see, e.g., Thomson (2012).

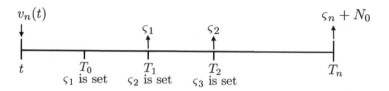

Fig. 4.5 Cash flow for a floating rate note.

We note that simple spot rates $F(T_{i-1}, T_i)$ (e.g., such market rates as LIBOR) are determined at time T_{i-1} (and hence in Definition 4.2 we need the time moment T_0) and, consequently, the coupon ς_i to be paid at time T_i is known at time T_{i-1}.

Let us find the arbitrage price $v_n(t)$ of a floating rate note at time $t \leq T_0$. It follows from (4.2) and (4.15) that

$$\varsigma_i = N_0 \left(\frac{1}{P(T_{i-1}, T_i)} - 1 \right). \tag{4.16}$$

We pay particular attention to the fact that the value of ς_i is not known until time $t = T_{i-1}$. Let us discover its value at the time t of $1/P(T_{i-1}, T_i)$ paid at time T_i, which, as usual, we do via a strategy:

- At t buy one bond with maturity T_{i-1} for $P(t, T_{i-1})$.
- At T_{i-1} receive \$1 according to the bought bond and use it to buy $1/P(T_{i-1}, T_i)$ bonds with maturity T_i.
- At T_i receive $\$1/P(T_{i-1}, T_i)$.

Hence, at time t the value of $\$1/P(T_{i-1}, T_i)$ paid at time T_i is equal to our initial investment $P(t, T_{i-1})$, otherwise – arbitrage. This results in the value of ς_i from (4.16) paid at time T_i being equal at time t to

$$N_0 \left[P(t, T_{i-1}) - P(t, T_i) \right]. \tag{4.17}$$

Summing up the cash flow, we obtain

$$v_n(t) = N_0 \sum_{i=1}^{n} \left[P(t, T_{i-1}) - P(t, T_i) \right] + N_0 P(t, T_n) = N_0 P(t, T_0), \tag{4.18}$$

which is a very simple formula! We also see that at the first reset date $v_n(T_0) = N_0$.

Example 4.6. Find the price of a floating rate note with a face value of \$100 that matures in three years, pays notes semiannually and the first note is paid in six months, given the current instantaneous forward rate curve as in Example 4.5.

Answer. According to (4.18), we have $v_6(0) = 100 \times P(0, 0.5)$ and according to (4.6):

$$v_6(0) = 100 \times \exp\left(-0.01 \int_0^{0.5} \ln(150 + 48u)du\right) \approx 97.49 \ (\$).$$

4.3 Interest-Rate Swaps

The Interest-Rate Swap (IRS) is a contract allowing the exchange of a payment stream at a fixed-interest rate for a payment stream at a floating rate (usually LIBOR plus some fixed margin).

Let us specify a number of dates $T_0 < T_1 < \cdots < T_n$ for simplicity equidistant with $\delta = T_{i+1} - T_i$ for $i = 1, \ldots, n-1$.

Definition 4.3. *A **payer (receiver) interest-rate swap** with a fixed rate ρ and a nominal value N_0 settled in arrears*[9] *is a contract according to which its holder pays (receives) fixed $\rho\delta N_0$ and receives (pays) floating $F(T_{i-1}, T_i)\delta N_0$ at the coupon dates T_i, $i = 1, \ldots, n$.*

The payer IRS is also called the *fixed leg* of a swap and the receiver IRS is also called the *floating leg* of a swap. We consider here plain vanilla IRS only, see some exotic IRS in, e.g., Kolb (2003).

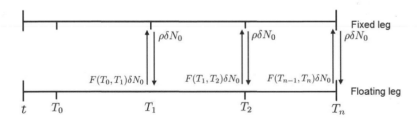

Fig. 4.6 Cash flow for an interest rate swap.

We note that cash flows take place only at the coupon dates. The net cash flow at time T_i, $i = 1, \ldots, n$, for a payer IRS is $[F(T_{i-1}, T_i) - \rho]\,\delta N_0$ and for a receiver IRS is $[\rho - F(T_{i-1}, T_i)]\,\delta N_0$. It follows from (4.15) and

---

[9]It means settled at the end of the contract.

(4.17) that the payment $F(T_{i-1}, T_i)\delta N_0$ to be received at time T costs now at time t:

$$(P(t, T_{i-1}) - P(t, T_i)) N_0.$$

Then, the value $\Pi_p(t)$ of the payer swap at time t is equal to

$$\Pi_p(t) = N_0 \sum_{i=1}^{n} [P(t, T_{i-1}) - P(t, T_i) - P(t, T_i)\rho\delta] \qquad (4.19)$$

$$= N_0 \left(P(t, T_0) - P(t, T_n) - \rho\delta \sum_{i=1}^{n} P(t, T_i) \right).$$

Note that we know this value at time t with certainty. It is obvious that the value $\Pi_r(t)$ of the receiver swap at time t is equal to $-\Pi_p(t)$.

A payer IRS can be viewed as a portfolio of an issued (fixed) coupon bond and the corresponding bought floating rate note, while a receiver IRS can be viewed as a portfolio of an issued floating rate note and the corresponding bought coupon bond (see Problem 33 in Chapter 6).

Now let us find the 'fair' fixed-interest rate ρ, which is called the **forward swap rate** (or *par swap rate*) $R_{swap}(t) = R_{swap}(t; T_0, T_n)$ at time $t \leq T_0$. This question is analogous to finding the forward price for forward contracts considered in Section 3.2. The fixed-interest rate ρ is 'fair' (i.e. at the time the contract is entered into, there is no advantage to either party) if the prices of both swap legs are the same, i.e. $\Pi_r(t) = \Pi_p(t) = 0$. Hence, the forward rate swap $R_{swap}(t; T_0, T_n)$ is equal to for $t \leq T_0$:

$$R_{swap}(t; T_0, T_n) = \frac{P(t, T_0) - P(t, T_n)}{\delta \sum_{i=1}^{n} P(t, T_i)}. \qquad (4.20)$$

We see that IRS does not require an up-front payment from either party like the forwards we considered in Section 3.2. The value of the swap varies with time, departing from the initial value of zero. During the life of the swap the same valuation technique is used, but, since over time the forward rates change, the value of the variable-rate part of the swap will deviate from the unchangeable fixed-rate side of the swap.

Exercise 4.5. Show that at time $t = T_{i-1}$, $i = 1, \ldots, n$, the value of the payer swap entered into at time T_0 equals

$$\Pi_p(t) = N_0\delta \left(R_{swap}(t; T_{i-1}, T_n) - R_{swap}(T_0; T_0, T_n) \right) \sum_{j=i}^{n} P(t, T_j). \qquad (4.21)$$

Table 4.2 The simple forward rates $F(0; T_i, T_j)$ in Example 4.7.

t	T	$F(0; t, T)$
0	0.5	0.05
0.5	1	0.08
1	1.5	0.09
1.5	2	0.08

Example 4.7. Let today be $t = 0$, the first reset date of a swap be $T_0 = 0.5$, the cash flow dates be at $T_i = (i + 1)/2$, $i > 0$ and the swap maturity time be $T_3 = 2$. The simple forward rates $F(0; T_i, T_j)$ are given in Table 4.2.

(i) Find the corresponding term structure of bond prices $P(0, T_i)$, $i = 0, \ldots, 3$.

(ii) Find the corresponding swap rate $R_{swap}(0; 0.5, 2)$.

Answer. This elementary problem is for your self-study. For your verification, the answers are (to four decimal places): $P(0, 0.5) = 0.9756$, $P(0, 1) = 0.9381$, $P(0, 1.5) = 0.8977$, $P(0, 2) = 0.8632$ and $R_{swap}(0; 0.5, 2) = 0.0833$.

Let us consider two examples of the use of IRS in practice. For other examples, see, e.g., Hull (2003) and Kolb (2003).

Example 4.8. (*Use of interest-rate swaps*). Consider two companies: REd which has an AAA credit rating[10] and BLue which has a lower A credit rating. Suppose that both wish to borrow \$100 million for five years and have been offered the rates as in Table 4.3.

Table 4.3 The rates used in Example 4.8[11].

	Fixed, % p.a.	Floating
REd	8.0	six-month LIBOR + 30 bp
BLue	9.2	six-month LIBOR + 100 bp

We also assume that BLue wants to borrow at a fixed rate of interest and REd wants to borrow at a floating interest rate linked to the six-month LIBOR. Because BLue has a worse credit rating than REd, it pays a higher interest rate than REd in both fixed and floating markets. In this example a possibility of a mutually beneficial IRS arises from the fact that the difference between the two fixed rates is greater than the difference between the two

---

[10]In this example a credit rating has the meaning of an evaluation by a credit rating agency of the company's ability to pay back their debt and the likelihood of their default. The lower the rating the higher risk of defaulting.

[11]Here bp means base point and 1 bp = 0.01%.

floating rates. Indeed, BLue pays 1.2% more than REd in fixed-rate markets while just 0.7% more than REd in floating-rate markets. Hence, BLue has a relative advantage in floating-rate markets and REd has a relative advantage in fixed-rate markets. We further assume that REd and BLue have directly contacted each other – this is a simplification since normally a financial intermediary is involved in swap contracts (see details, e.g., in Hull, 2003; Kolb, 2003). Consider the following IRS between REd and BLue: REd agrees to pay BLue interest at the six-month LIBOR flat rate on $100 million on the floating leg of the IRS and BLue agrees to pay REd interest at the fixed rate of 7.95% p.a. on $100 million on the fixed leg of the IRS. So, REd has the following sets of interest-rate cash flows (see also Fig. 4.7):

- It borrowed on the fixed-rate market taking the comparative advantage: it pays 8% p.a. to outside lenders.
- It receives 7.95% p.a. from BLue according to the IRS.
- It pays LIBOR to BLue according to the IRS.

The net effect is that REd pays LIBOR plus 5 bp on the borrowed $100 million required for its business. This is 0.25% p.a. less than it would pay if it went directly to the floating-rate market.

BLue has the following cash flows (see also Fig. 4.7):

- It borrowed on the floating-rate market taking the comparative advantage: it pays LIBOR + 100 bp p.a. to outside lenders.
- It receives LIBOR from REd according to the IRS.
- It pays 7.95% p.a. to REd according to the IRS.

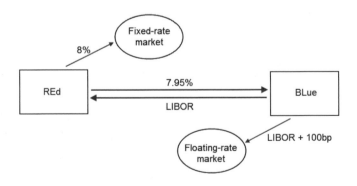

Fig. 4.7   Illustration for Example 4.8 on the use of IRS.

The net effect is that BLue pays 8.95% p.a. on the borrowed $100 million required for its business which is 0.25% p.a. (i.e. 25 bp) less than it would pay if it went directly to fixed-rate markets. We see that they entered in the IRS which is clearly beneficial for both parties.

Remark 4.2. At the first glance, it looks like REd and BLue in the above example exploited market informational inefficiency (arbitrage). But modern swap markets (in comparison with the early days of 1980s) are mature, highly liquid and well understood. Therefore, to find arbitrage opportunities on the markets today is virtually impossible. To understand why the above example nevertheless has modern practical importance, we need to look further into the logic of financial markets. The reason for the spread differentials is hidden in how contracts available to the companies in fixed and floating markets work. Note that the fixed rates in the above example are five-year rates while LIBOR rates are six-month rates. In the floating-rate market, the lender can usually review the floating rates every six months. If the credit rating of REd or BLue is cut, the lender may increase the spread over LIBOR or it can even refuse to roll over the loan; however, fixed-rate lenders cannot change the terms of the loan. The spreads between the rates offered to REd and BLue reflect the probability that BLue is more likely to default than REd. During the next six months, the chance of either REd or BLue defaulting is very low. However, it is statistically known that the probability of a default by a company with a lower credit rating increases faster than for a company with a higher credit rating. This is the reason for the spread between the five-year rates being greater than the spread between the six-month rates. This also means that the IRS deal in the above example carries credit risk[12] and the deal is beneficial only if a default does not happen, i.e. the observed benefits are not a 'free lunch'.

Example 4.9. (*Use of interest-rate swaps*). Consider a savings and loan association. It takes deposits and lends the received money for long-term mortgages. Most deposits are placed for short periods of time and hence interest rates on them must adjust to changing interest rates on the market. At the same time, many mortgages are borrowed at a fixed rate for long time periods. Therefore, the savings and loan association has floating-rate liabilities (they have to pay floating rates) and fixed-rate assets (they receive a fixed

---

[12]Credit risk is the risk of loss of a principal or a financial reward by an investor due to a borrower's failure to make payments according to the contract. We do not consider credit risk in this course. Some introductory reading on credit risk can include Hull (2003) and for comprehensive coverage see, e.g., Bielecki and Rutkowski (2010) and Schmid (2004).

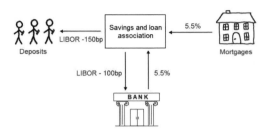

Fig. 4.8   Illustration for Example 4.9 on the use of IRS.

rate). Hence, for the association there is a risk due to possible rise of variable rates. Indeed, if floating rates rise, the association will have to increase the rate it pays on deposits (if it does not do so then the customers will move deposits to a different bank which pays higher rates). However, it cannot increase mortgage rates to match the loss on the deposits side because rates on the existing mortgages are fixed. To avoid this risk, the association can use the swaps market: either transform its fixed-rate assets into floating rate assets (i.e. enter into a payer IRS) or transform its floating-rate liabilities into fixed-rate liabilities (i.e. also via a payer IRS). To clarify the matter, let us look at an example with numbers. Suppose that the association has a mortgage book of £10 million at the fixed rate of 5.5% for the period of ten years. It also guarantees to pay a LIBOR minus 150 bp on deposits. A financial institution offers an IRS with a fixed rate of 5.5% and floating rate LIBOR minus 100 bp. The association enters in this IRS (its fixed leg). As a result (see also Fig. 4.8), it receives mortgage payments of 5.5% on the principal of £10 million and passes them to the financial institution under the IRS, for which the association receives a floating-rate LIBOR minus 100 bp. From this cash flow, the association pays its depositors LIBOR minus 150 bp. Hence it will have a periodic inflow of 0.5%, which is the spread it makes on the loan and it now avoids the interest-rate risk, i.e. no matter what happens with interest rates over the next ten years, the association will receive a net cash inflow of 0.5% on £10 million[13].

---

[13]We note in passing that the association still has the risk of default by mortgage holders or by its counterparty in the swap deal but this is a different story for a different course.

Chapter 5

European Options

In this chapter we introduce another type of derivative – options. An option is a possibility or a choice, e.g., you have an option to stop reading this book right now and sell it. Here we will consider options that are traded on financial markets[1]. Recall that futures/forwards are contracts that bind both parties. Contrary to that, options are contracts in which only one partner assumes the obligation, whereas the other obtains a right.

Definition 5.1. *An **option** is a contract which gives the right but not the obligation to do something at a future date.*

There are many different kinds of options. The simplest type of options is a European one. We start by considering European *plain vanilla options* – European calls and puts.

Definition 5.2. *A (plain vanilla) **European option** is a contract between two parties in which:*

*The seller of the option (the **writer**) grants the buyer of the option (the **holder**) the right to purchase (**call**) from the writer or to sell (**put**) to him an underlier with current spot price S_t for a prescribed price K (**exercise or strike price**) at the **expiry date (maturity time)** T in the future.*

---

[1]Note that we will consider options which work for equity and commodity markets, however, we will not consider options of interest-rate (in other words, fixed-income) markets like swaptions, caps, floors, etc. Though their nature is, in principle, similar to the options we consider here, they have their own specifics (essentially, because of the term structure). The corresponding material can be found, in e.g., Avellaneda and Laurence (2000); Brigo and Mercurio (2006); Filipovic (2009); Björk (2004) and Schoenmakers (2005).

The key property of an option is that only the writer has an obligation.

Example 5.1. (See also Kindleberger, 2000). Historically, the first major use of option-like contracts occurred in the Netherlands in the 17th century. At the time tulips were very popular. Tulip growers wanted to protect themselves against market price fluctuations. Therefore, they purchased contracts which entitled them to sell tulip bulbs at a minimum price K if the market price S_T decreased below the threshold. The writers of the contracts expected the prices to increase and hence that the growers would not exercise their right. The writers thought the premiums charged for these options would be an easy profit. However, the tulip market collapsed in 1637. The economic impact was huge and the writers of the put options were unable to honour their obligation to buy. After this event, options in Europe were not popular for a long time. They have been traded in London since the 18th century but it was not until the 1970s that options became popular (recall the reasons for that given in Chapter 1).

Example 5.2. When we discussed the farmer and the miller in previous examples, they used a forward contract to reduce risk. This can also be done using a more sophisticated financial instrument – an option. For the farmer an option can be used as an insurance. He can estimate the minimum price K that he can accept for his business to remain intact. Then he goes to a financial institution (a writer) and buys a European put option for his crop: the farmer is ready to pay a small premium to the writer for the right to sell  his crop at the price K. In Autumn (at the maturity time T) if the stock price $S_T < K$, the farmer will go to the writer and exercise the option. As a result, he will get the difference

$$K - S_T.$$

If $S_T > K$, the farmer will obviously not exercise his right and will sell his crop on a market for the price S_T.

Thus, the *payoff* (payoff function) of the European put option at time T (i.e. how much the writer shall pay to the holder) is

$$f(S_T) = (K - S_T)_+ := \max(K - S_T, 0).$$

As an illustration of a European call option, consider the miller who may wish to secure himself from a rise in prices in grain. The payoff of a European call option at time T is

$$f(S_T) = (S_T - K)_+.$$

The payoff diagram for a put option is plotted in Fig. 5.1 and the payoff diagram for a call option is plotted in Fig. 5.2.

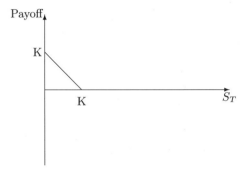

Fig. 5.1   The payoff of a European put option.

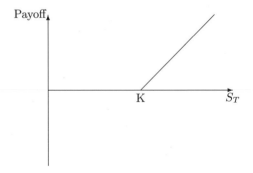

Fig. 5.2   The payoff of a European call option.

5.1   Moneyness

Let us introduce some additional market terminology. *Moneyness* is attributed to the profit or loss which could occur from the immediate exercise of an option.

A call (or put) is *in-the-money* (ITM) if the stock price exceeds (is below) the strike price. For example, a call option with the strike of £100 and the underlying stock price of £105 is £5 ITM.

A call (put) option is *out-of-the-money* (OTM) if the stock price is less than (exceeds) the strike price. For example, a put option with the strike of £100 and on the stock with the spot price of £105 is £5 OTM.

A call or put option is *at-the-money* (ATM) if the stock price equals (or is very near to) the exercise price. One can also say that a put or a call is *deep-in-the-money* (*deep-out-of-the-money*) if they judge the profit (loss) from an immediate exercise of the option to be too large.

5.2   Reading Option Prices

Table 5.1 shows an example of quotations for put and call options (of any kind, not necessarily European) from a newspaper. Prices are from 1 October this year for trading options on the stock XYZ. Options expire in November and December[2] this year and have strikes 90, 95, 100 and 105 ($).

Table 5.1   Example of options' quotations.

XYZ	K	c Nov	c Dec	p Nov	p Dec
$97\frac{7}{8}$	90	s	$9\frac{1}{8}$	s	$1\frac{3}{8}$
	95	$4\frac{3}{8}$	$5\frac{1}{2}$	$2\frac{5}{8}$	$3\frac{1}{8}$
	100	$1\frac{7}{8}$	$3\frac{1}{8}$	r	$6\frac{1}{4}$
	105	$\frac{13}{16}$	$1\frac{3}{4}$	r	10

In Table 5.1, 'r' means that the option is not traded on the day and 's' that this option is not listed for trading. These letters have no specific meaning and are merely the convention used by *The Wall Street Journal*.

Consider the call with strike $100 that expires in November. This option has a price of $1\frac{7}{8}$ or $1.875. This is the *trading price* or *premium price* or, simply, *premium* of the option. In the case of the option (unlike the forward) we pay a premium to enter the contract. Notice that the premium is much less than the strike and the spot price of the underlying stock. The price given in the table is the price of a put or call for a single share. Option contracts are often written for 100 shares. To buy such a contract with a strike price of $100 and maturity in November would cost $187.50. Assuming this is a

---

[2]In the UK the expiry date is 18:00 of the third Wednesday of the month.

European-type call, owning it would give the holder the right to purchase 100 shares of XYZ at $100 per share at the expiry.

From looking at Table 5.1 we can make some further observations.

1. Option prices are usually higher the longer the time until the option expires, which corresponds to our intuition because more time means more uncertainty.
2. For a call, a lower strike means the option is more valuable. Indeed, the strike is what the call holder pays for the stock. The lower the amount he pays, the better the financial product.
3. For a put, a higher strike increases the price of the option.

See also Section 5.6, where we discuss further properties of European plain vanilla options.

5.3 Profit and Loss

We have seen payoff diagrams for call and put options. Now we will look at the profits and losses of writers and holders after taking into account the premium. Profit from an option is

$$\text{profit} = \text{payoff} - \text{premium}.$$

A holder of a European call option with strike price K and price C has limited loss but unlimited gain (see the profit/loss diagram in Fig. 5.3).

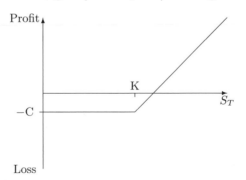

Fig. 5.3   Profit/loss diagram for a holder of a European call.

A writer of the same call option has unlimited loss but only a limited gain (see the profit/loss diagram in Fig. 5.4).

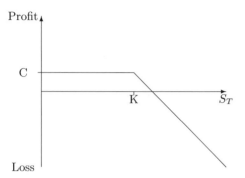

Fig. 5.4   Profit/loss diagram for a writer of a European call.

A holder of a European put option with strike price K and price D has limited gain and limited loss (see the profit/loss diagram in Fig. 5.5). Note that in the literature one usually denotes a put price by P, not D. We use D in this course to avoid confusion with the notation for probability.

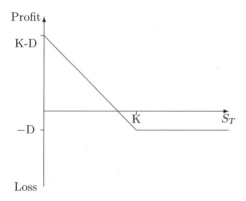

Fig. 5.5   Profit/loss diagram for a holder of a European put.

A writer of a put option has limited loss and limited gain (see the profit/loss diagram in Fig. 5.6).

These diagrams illustrate that the option market is a zero-sum game, i.e. the option's holder profit is the writer's loss and *vice versa* (ignoring taxes and transaction costs).

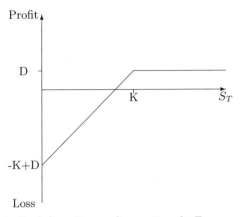

Fig. 5.6   Profit/loss diagram for a writer of a European put.

5.4   Why Buy Options?

Many people use options, but why do they choose them over other financial instruments? Let us consider a couple of examples.

Example 5.3. A dealer is buying some assets (long position) and wishes to reduce his possible losses due to decreasing prices. To this end he may buy a put option. This will:

- provide a cheap way of eliminating possible losses;
- but reduce possible profit.

For instance, given $r = 0$ (no interest), $S_0 = \$100$ (current stock price), $K = \$100$ (strike price) and $D = \$2$ (price of put) then if:

$S_T = \$20$		Profit/loss
	without put	$-\$80$
	with put	$-\$2$
$S_T = \$120$		
	without put	$+\$20$
	with put	$+\$18$

Example 5.4. A dealer is borrowing some asset (or undertaking the delivery in question, i.e. he has the short position) and may wish to reduce his possible loss from an increase of the price of his stock. Then he might buy a call option which will:

- eliminate risk in case of stock price increase;
- but reduce possible profit.

Exercise 5.1. Example with numbers. Please make it up yourselves.

There are two reasons to buy an option:

1. Hedging (see Examples 5.2–5.4).
2. Speculation.

There are chances to win a profit spending a small amount of money (called *leverage* or *gearing*), much less than buying or selling the underlying stock. Remember that an option price is much less than its underlier's price.

We have already considered a number of examples illustrating the use of options for hedging. Let us now look at how they can be used for speculation.

Example 5.5. (*'Bulls'*). You win £1,000 in a lottery and are wondering what to do with the money. You love a particular company and strongly believe that their shares will go up. We assume $r = 0$ to neglect the time value of money. Suppose that $S_0 = £10$, you buy 100 shares. Then at some time T in future:

- you could lose up to £1,000;
- if $S_T = £11$, your profit would be £100, i.e. 10%.

Alternatively, suppose you can buy a call option with $K = £10$, maturity T and price $C = £0.10$. You buy this option on 110 shares, so you spend £11 out of your £1,000. Then at time T:

- you could lose a maximum of £11 (the premium);
- if $S_T = £11$, you get a profit of £110−£11 =£99, i.e. 900%! You also have the remaining £989 from your win.

Note, however, that if $S_T < £10$ even by one pence, you lose £11 while shares can recover.

Example 5.6. (*'Bears'*). A different person wins £1,000 in a lottery and has a firm belief that a particular company is 'bad' and that its shares will fall in value. Assume that the current spot price of its share is $S_0 = £100$ and the price of the put with strike $K = £100$ and maturity T written on the company's single share is $D = £1$. Further, suppose they buy a put for 11 shares. Then, at time T:

- maximum loss is £11;
- if $S_T = £90$, the profit is £99.

As for the writer (of both call and put), he may hope that market will move in the right direction for him and he will keep the premium (see also Example 5.1).

There are special names for those acting on the assumption of a rise or fall of some asset. The dealers expecting prices to go up are called *bulls*.

'...Bull markets are born on pessimism, grow on scepticism, mature on optimism and die on euphoria...' – Sir John Templeton (http://www.sirjohntempleton.org/quotes.asp [Accessed on 26 April 2013]).

A bull opens a long position expecting to sell with profit afterwards when the market goes up. Dealers who expect the market to move downwards are called *bears*. A bear tends to sell securities he has (or even has not – by short selling). He hopes to close his short position by buying the traded items at lower prices at a later time.

Futures and options are similar instruments for speculators since both financial instruments give an opportunity to obtain leverage. However, there is an important difference between the two. With futures, the speculator's potential loss, as well as the potential gain, is very large. With options, the speculator's loss is limited to the amount paid for the option.

5.5   Put–Call Parity

The key question is how to price options. What is a fair price for both holder and writer? Options are more complicated than forwards, so how do we price them? Do they have an arbitrage price? We will address these questions in the near future but firstly consider another feature of European options. There is an important relationship between the prices of European puts and calls with the same maturity, strike and underlying.

Let C_t and D_t be the prices of a European call and put, respectively, at time t with strike price K and maturity time T on a stock of price S_t; let B_t be the amount of money in a bank account at time t and r be the continuously compounded interest rate. We assume that the stock is not paying dividends (see also Problem 30 in Chapter 6). Consider two portfolios:

1. $V_t^1 = C_t + B_t$, which becomes at $t = T$ (see also Fig. 5.7):

$$V_T^1 = (S_T - K)_+ + e^{r(T-t)}B_t = \begin{cases} e^{r(T-t)}B_t & \text{if } S_T < K, \\ S_T - K + e^{r(T-t)}B_t, & \text{otherwise.} \end{cases}$$

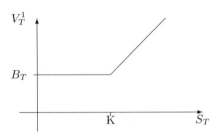

Fig. 5.7   Dependence of the value of the first portfolio on the spot price at the option's maturity.

2. $V_t^2 = D_t + S_t$, which becomes at $t = T$ (see also Fig. 5.8):

$$V_T^2 = (K - S_T)_+ + S_T = \begin{cases} K & \text{if } S_T < K, \\ S_T, & \text{otherwise.} \end{cases}$$

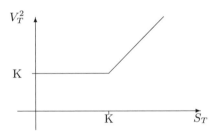

Fig. 5.8   Dependence of the value of the second portfolio on the spot price at the option's maturity.

If $B_T = K$ (i.e. $B_t = Ke^{-r(T-t)}$), then

$$V_T^1 = V_T^2,$$

independently of any randomness. Thus, these 'stock plus put' and 'bank account plus call' combinations have the same payoffs in all possible future states of the world.

As agreed before, we are assuming that there are no arbitrage opportunities. This implies that the current values of the portfolios should be the same as well. So

$$C_t + Ke^{-r(T-t)} = D_t + S_t$$

or

$$C_t = D_t + S_t - Ke^{-r(T-t)}.$$

This is the **put–call parity relationship** for European options on the underlying that pays no dividends.

We note that in deriving the put–call parity we did not assume any model for the price behaviour S_t, we used the no-arbitrage principle only. If the put–call parity does not hold, we can find a strategy which will give us a risk-free profit. This is illustrated in the following example.

Example 5.7. Let S_0 =£40, r = 0.05 and prices of call C_0 = £3 and put D_0 = £2 with the same strike K = £40 and maturity T = 1/3 (i.e. four months). As the call is selling at £3, we know from the put–call parity relationship that the put should be worth

$$D_0 = Ke^{-rT} - S_0 + C_0 \approx £2.34.$$

This suggests that the quoted put with D_0 =£2 is underpriced relative to the call or, similarly, the call is overpriced. Then we can take advantage and get a risk-free profit by using the strategy described in Table 5.2 (the logic here is to buy what is underpriced and to sell what is overpriced).

Table 5.2 The strategy in Example 5.7.

$t = 0$		$T = 1/3$	
		$S_T \leq 40$	$S_T > 40$
Write call	£3	0	$40 - S_T$
Buy one put	−£2	$40 - S_T$	0
Buy one stock unit	−£40	S_T	S_T
Borrow	£39	−£39.655	−£39.655
Total value of portfolio	£0	£0.345	£0.345

This was about just one share. If we expand this strategy to, e.g., ten option contracts for 100 shares each on both sides, our profit becomes £345.

We conclude that if we can value a European call option, we will value a European put option as well using the put–call parity.

5.6   Basic Properties of European Calls and Puts

Let us discuss some properties of European calls and puts. As we have already seen in previous sections, their current prices C_0 and D_0 depend on the following parameters: strike K, time to the maturity T, spot price of the stock S_0 and interest rate r. They also depend on dividend yield and on the volatility of the stock. *Volatility* σ we understand to be a measure for variation in the stock price (in the following chapters we will be more specific and define it as standard deviation of the stock log-return). Roughly speaking, we can say that the higher the volatility the more uncertain the behaviour of the stock. These six are the main parameters which influence prices of options. Other parameters (transaction costs, taxation, expected growth of stock price, attitude towards risk by investors) usually only have a minor influence on option prices (Cox and Rubinstein, 1985).

As before, for simplicity, we restrict ourselves to stocks that do not pay dividend. We will look at the effect of the other five main parameters on option prices and in doing so we will assume that only one of these parameters increases while the other four remain fixed. This assumption is obviously not valid in practice, e.g., when the interest rate decreases, spot stock prices usually increase.

If the strike K increases then the value of a call decreases while the value of a put increases (see discussion in Section 5.2 and also Problem 22 in Chapter 6).

Buying an option, we are protecting ourselves against uncertainty. As we already discussed in Section 5.2, the longer maturity time means more uncertainty and hence heuristically we should pay more for the protection, i.e. prices of both calls and puts increase with the increase of maturity time. By the same argument, increase in volatility leads to the increase of option prices.

The higher the spot price S_0 the more expensive the call and the cheaper the put. Heuristically, if we start at a larger level of S_0, we are more likely to reach a higher price S_T at the maturity (recall that all the other parameters are assumed to be fixed) implying that the payoff of the call $(S_T - K)_+$ is likely to be larger and the payoff of the put $(K - S_T)_+$ is likely to be smaller.

An increase of the interest rate r decreases the present value of any future cash flow. When interest rates increase, it is expected that the economy grows faster and, consequently, investors expect the stock price to increase. Both factors lead to a decrease of the price of a put. For a call, the first factor decreases a call's value while the second one increases it. However, it can be shown that the call option price increases with the increase of the interest rate r. Heuristically, we can explain this by noting that one usually expects that the discounted stock price $\tilde{S}_T = e^{-rT}S_T$ remains unchanged when r changes. Because the present value of a call is equal to $e^{-rT}(S_T - K)_+ = (\tilde{S}_T - e^{-rT}K)_+$, an increase of the interest rate leads to the effective decrease of the strike and hence to the increase of the call's price.

Table 5.3 Propeties of European call and put options.

Increase of	Price of call	Price of put
Strike	Decrease	Increase
Maturity	Increase	Increase
Spot stock price	Increase	Decrease
Volatility	Increase	Increase
Interest rate	Increase	Decrease

For convenience, the main properties of European plain vanilla options which have been discussed are summarised in Table 5.3. You can read more on properties of options and their institutional settings in, e.g., Hull (2003); Cox and Rubinstein (1985); Kolb (2003) and Bingham and Kiesel (2004). We will return to the properties of options in the last part of the course where we will consider sensitivities (or so-called Greeks) of option prices.

We have finished the first part of the course, in which we introduced and illustrated the no-arbitrage principle as well as considered such financial instruments as forwards, futures, bonds, swaps and options. In the next part our main objective is to answer the question of pricing and hedging of options, for which we will need to revise and extend the knowledge of Probability theory beyond assumed prerequisites for this course.

Chapter 6

Problems for Part I

1. You take a mortgage of £200,000 on 1 February 2013 at a fixed-interest rate of 5% p.a. with continuous compounding (consider each month as 1/12 of a year). It is repayable over 25 years with level payments due at the end of every month (your last payment is on 31 January 2038). Calculate the monthly payment amount.

2. An investor put £100,000 into a savings bank account with a fixed interest rate. In return she will receive £111,529 in five years. Calculate the interest rate p.a. with
   (a) annual compounding;
   (b) monthly compounding;
   (c) continuous compounding.

3. Exactly five years ago you invested £X in a company which promised to pay you a constant rate of continuously compounded interest of 8% p.a. as long as you do not withdraw the funds. Suppose that you still keep this investment and its current value is £50,000.
   (a) Find X.
   (b) If you keep this investment for another five years, what will be its value then?

4. A bank quotes you an interest rate of 3.5% p.a. with quarterly compounding. What is the equivalent rate with
   (a) daily compounding;
   (b) monthly compounding;
   (c) continuous compounding.

5. A 12-month long forward contract on 100 tonnes of feed wheat is entered into when the commodity price is £145.05 *per tonne* and the interest rate is 0.5% p.a. with continuous compounding. (Neglect costs of storage and convenience yields.)

(a) What is the forward price for the contract at the date of entry?

(b) Six months later, there is a shortage of feed wheat on the market and its price jumps to £151.96 *per tonne*. The cost of borrowing also increases and the interest rate becomes 1.25%. What is the new forward price (for the remaining six months) and what is the current value of the original long forward contract?

(c) Suppose one agrees to enter into a 12-month forward contract on 100 tonnes of feed wheat with the forward price £14,500 when the wheat price is £145.05 per tonne and the interest rate is 0.5% p.a. with continuous compounding. What arbitrage opportunities does this create? Demonstrate them via an appropriate strategy.

(d) Now suppose that costs of storage and convenience yield should be taken into account. How do these new conditions change your answer to part (a) of this question if we assume that the present value of all the storage costs per unit of this commodity that will be incurred during the life of the forward contract is £0.25 per tonne and the convenience yield is 0.25% p.a.?

6. The risk-free fixed rate of interest is 2% p.a. with continuous compounding and the dividend yield on a share of an XYZ company is 4% p.a. with continuous compounding (i.e. the asset pays a continuously compounded interest on its instant price). The current value of the share is £100. What is the one-year forward price for one share of this type?

7. Suppose the fixed-interest rates in the UK and Eurozone are 2% and 4% p.a., respectively, with continuous compounding. Also, assume that in the Euro market the current Sterling exchange rate is €1.200 (so that £1,000 costs €1,200). What is the fair futures price in the Eurozone for a contract on £1,000 deliverable in six months? (Assume that a futures contract is the same as a forward.)

8. Explain the difference between a long and short forward.

9. An 18-month short forward contract on 1,000 barrels of crude oil is entered into when the commodity price is $60.4 *per barrel*[1] and the interest rate is 2% p.a. with continuous compounding. (Neglect costs of storage and convenience yields.)

(a) What is the forward price for the contract at the date of entry?

(b) Six months later, the price of oil is dropped to $47.1 *per barrel* (the interest rate is unchanged). What is the new forward price (for the

---

[1] This was price of oil in Autumn 2009.

remaining 12 months) and what is the current value of the original short forward contract?

(c) Suppose one agrees to enter into an 18-month forward contract on 1,000 barrels of crude oil with the forward price $63,000 when the price is $60.4 *per barrel* and the interest rate is 2% p.a. with continuous compounding. What arbitrage opportunities does this create? Demonstrate them via an appropriate strategy.

10. *Mark-to-market* (also known as fair value accounting) refers to the accounting act of recording the value of an asset or liability according to its (or similar assets and liabilities) current market value rather than its book value[2].

Suppose Investor X entered into a long forward contract on 100 shares of ABC for 12 months as OTC, with no regular margins to be paid. (Assume that there is no initial margin[3] and maintenance margin[4].)

Suppose Investor Y entered into a long futures contract also on 100 shares of ABC for 12 months, according to which margins were to be paid/received every three months in cash, i.e. the difference between the futures' mark-to-market and book values. Her profits/losses arising due to these adjustments are rolled over at the three-month interest rate.

The observed performance of the ABC shares and the continuously compounded interest rate r_t over these 12 months are recorded in Table 6.1.

Table 6.1 The observed performance of the ABC shares and the continuously compounded interest rate r_t.

Time in months	ABC, £ per share	r_t, p.a.
0	0.97	3.5%
3	0.92	3%
6	0.95	2.75%
9	1.01	2.0%
12	1.2	2.0%

Compute the profit/loss of Investors X and Y at the end of 12 months.

11. The current price of cocoa is $3,136 *per tonne*. Forward contracts are available to buy or sell 50 tonnes of cocoa at $170,000 with delivery in one year. Money can be borrowed at 3% p.a. What should be the arbitrageur strategy? (Neglect costs of storage and convenience yields.)

---

[2]For further reading, see e.g., Kolb (2003).

[3]It is a margin which must be deposited in the margin account at the exchange at the start of the contract.

[4]It is the minimum balance that is required to be maintained in the margin account during the life of the contract.

12. What is the difference between a forward and an option?

13. Suppose an XYZ stock is selling for $125 per share in February and an XYZ European call with the strike price $130 and expiry in August is selling for $10. Assuming a continuously compounded interest rate of 3% per year,
    (i) What is the value of the corresponding put?
    (ii) If one agrees to buy a put for $14, what arbitrage opportunities does this create? Demonstrate them via an appropriate strategy.

14. Give a possible reason why an *investor* might purchase a *call* option.

15. Let $0 < K_1 < K_2$. At its maturity T, a bear spread pays $\$A = K_2 - K_1$ if the spot price S_T of the underlying is less than K_1, it pays $K_2 - S_T$ if S_T is between K_1 and K_2, otherwise it expires worthless. How do you synthesise this derivative (i.e. how do you recreate its payoff) using plain vanilla put options?

16. How can a long forward contract on a stock with a particular delivery price and delivery date be created from European options?

17. Show that the values of European put and call with the same strike, maturity and underlying are equal if and only if the strike is equal to the forward price.

18. At time $t = 0$ an investor bought (long position) one European call with strike $K_1 = \pounds20$ and one European call with strike $K_2 = \pounds40$, she also borrowed (short position) two European calls with strike $K_3 = \pounds30$. All the options are on the same underlying and with the same maturity T. Such a portfolio is called a *butterfly spread*. Draw the payoff diagram for this butterfly spread.

19. Draw the payoff diagram at maturity of a *bull spread* with a long position in a call with strike $\pounds30$ and a short position in a call with strike $\pounds35$ (both calls are on the same underlying and with the same maturity).

20. Use arbitrage arguments to confirm the following inequality for European call option:
$$C_t \le S_t, \quad \text{for all } t \le T.$$

21. Assuming that the interest rate r is constant, prove the following inequality for a European call price C_0 at time $t = 0$ with strike K and maturity T:
$$S_0 - Ke^{-rT} \le C_0,$$
where S_0 is the price of the underlier at $t = 0$.

22. If two otherwise identical European call options with values $C_t^{K_1}$ and $C_t^{K_2}$ have exercise prices K_1 and K_2, respectively, and $K_1 < K_2$, then confirm by arbitrage arguments that $0 \le C_t^{K_1} - C_t^{K_2}$ for all $t \le T$.

23. Let the interest rate be zero. Consider a European put option with exercise price K and maturity time T on the underlier with the price S_t at time t. Denote by D_t the value of this put at time t. Confirm by arbitrage arguments that $D_t \geq (K - S_t)_+$ for all $t \leq T$.

24. (*Chooser option*). Let time moments T and $s \in [0, T]$, a strike price $K > 0$ and a continuously compounded interest rate r be given. A chooser option is a contract sold at time $t = 0$ that gives its holder the right to choose at the time $t = s$ either a call or a put which both expire at time $t = T$ and have the strike K. Using the put–call parity relation, show that at time $t = 0$ the price of a chooser option is the sum of the price of a put at time $t = 0$, which expires at time T and has the strike K, and the price of a call at time $t = 0$, which expires at time $t = s$ and has the strike $e^{-(T-s)r}K$.

25. Today is 10 August and the spot price of an XYZ share is £5. Also, European calls and puts on XYZ with strikes £3.5, £4, £4.5, £4.75, £5, £5.25, £5.5 and with maturities in September, November this year and January next year are now sold on the market. The XYZ's executive board will meet on 7 October this year to discuss the possible acquisition of another firm. You believe that, following this meeting, the XYZ share price will either go up or down by about 10% and will stay at that level for at least a month. Which portfolio of options would perform well if your guess were correct? (Neglect any transaction costs.)

26. What is the value of a European put option with strike $K = 0$?

27. In a real market, bid and ask prices[5] differ, i.e. there is a bid–ask spread. Suppose the bid and ask prices for a one-year European call with strike £30 on a non-dividend paying underlier with spot price £35 are £6 and £7, respectively, and the bid and ask prices for a one-year European put with the same strike and on the same underlier are £1 and £2, respectively. Assume that the risk free interest rate is 1% p.a. with continuous compounding. Is there an arbitrage opportunity present? Provide a full justification of your answer.

28. You expect that due to a reduction of your national government spending on new building programmes there is a large chance of a significant decrease of share prices of a big cement production company within the next two years.

---

[5]The bid price is a price that buyers are willing to buy at and the ask price is a price that sellers are willing to sell at. The difference between the bid and ask prices is called the bid–ask spread, which depends on the liquidity of the instrument and on the transaction cost.

(i) Which financial instruments will you recommend to a pension fund owning a large amount of shares of this company to use in order to protect their business from this risk?

(ii) If you speculate on the market, which financial derivatives would you use and how would you use them to try to profit from your prediction?

29. Call options with strikes $100, $110 and $120 on the same underlying asset and with the same maturity are traded for $4, $3 and $1, respectively. Is there an arbitrage opportunity present? If yes, how you can make a riskless profit?

30. Derive the put–call parity relationship for European options on the underlying stock which pays dividends, assuming that at time t the value of dividends to be paid over the period $[t, T]$ is known and given by G_t (suppose that the other assumptions from Section 5.5 hold).

31. Consider two portfolios. Portfolio 1: long two calls with strike equal to some $K > 0$. Portfolio 2: long one call with strike equal to $K - 10$ and long one call with strike $K + 10$. Assume that all the calls are on the same underlier and have the same maturity. Which of these two portfolios would you prefer to own? Provide a full justification of your answer.

32. Explain the difference between coupon bonds and floating rate notes.

33. Replicate a payer IRS using a coupon bond and a floating rate note. Then obtain the IRS price (4.19) as a result of this replication.

34. Let today be $s = 0$, the first reset date of a swap be $T_0 = 1/6$ (i.e. in two months time), the cash flow dates be at $T_i = T_{i-1} + 1/4$, $i > 0$, (i.e. every three months) and the swap maturity time be $T_4 = 7/6$. At time $s = 0$ the instantaneous forward rate is given by

$$f(0, t) = 0.05 + 0.01t.$$

(i) Find the corresponding term structure of bond prices $P(0, T_i)$, $i = 0, \ldots, 4$.

(ii) Find the corresponding swap rate $R_{swap}(0; 1/6, 7/6)$.

(iii) Suppose that after one year (i.e. at time $s = 1$) the instantaneous forward rate becomes $f(1, t) = 0.05 + 0.02t$ and the simple forward rate was $F(11/12; 11/12, 7/6) = 0.05$. What is then the price of the original receiver swap at the time 1? Assume that the nominal value is equal to £10 million.

35. Let today be $s = 0$, the first reset date of a swap be $T_0 = 1/12$ (i.e. in one month time), the cash flow dates be at $T_i = T_{i-1} + 1/12$, $i > 0$, (i.e. every month), and the swap maturity time be $T_4 = 5/12$. The simple forward rates $F(0; T_i, T_j)$ are given in Table 6.2.

Table 6.2 The simple forward rates $F(0; T_i, T_j)$.

t	T	$F(0; t, T)$
0	1/12	0.05
1/12	1/6	0.05
1/6	1/4	0.06
1/4	1/3	0.07
1/3	5/12	0.09

(a) Find the corresponding term structure of bond prices $P(0, T_i)$, $i = 0, \ldots, 4$.

(b) Find the corresponding swap rate $R_{swap}(0; 1/12, 5/12)$.

(c) Suppose that later the simple forward rates become $F(1/6; 1/6, 1/4) = 0.04$, $F(1/5; 1/5, 1/4) = 0.05$, $F(1/5; 1/4, 1/3) = 0.06$ and $F(1/5; 1/3, 5/12) = 0.06$. What is then the price of the original receiver swap at the time $1/5$? Assume that the nominal value is equal to £100 million.

36. Find the value of the payer swap $\Pi_p(t)$ at $t \in [T_{i-1}, T_i)$, $i = 1, \ldots, n$, entered into at time $s \leq T_0$.

Part II

Discrete-Time Stochastic Modelling
and Option Pricing

It's good to look ahead,
But first you need to know
The proper initial count:
One, two, three, four, five[1].

In this part of the course we deal with the pricing and hedging of European-type and American-type options in the discrete-time setting and also discuss complete and incomplete markets. Mathematics-wise, we briefly revise elementary Probability and then consider discrete-time discrete-space stochastic processes which will be used in this part for financial modelling.

---

[1]Translated from a song about Alice and digits from Visotsky's audioplay; see Footnote 1 on page 5.

Chapter 7

Binary Model of Price Evolution

To price a forward, we did not need a model for price evolution; however, to price an option, we need such a model. Here we consider a simple discrete-time discrete-space (by space, we understand the price of the underlying) model of price evolution. Its simplicity allows us to make the presentation of the material transparent. At the same time, this simple model is often used by financial institutions in pricing options, especially if we note that the tree model we are going to introduce can be viewed as a discretization of continuous models of price evolution described by stochastic differential equations (see, e.g., Glasserman, 2003; Milstein and Tretyakov, 2004).

Consider the equally spaced discrete times: $0 = t_0 < t_1 < \cdots < t_N = T$, $\Delta t := t_{n+1} - t_n$. We assume that there is only one type of stock you can own and one type of a bank account for which value $B_n \equiv B_{t_n}$ at time t_n is deterministic $B_n = B_0 e^{rt_n} = B_0 e^{rn\Delta t}$ with a known fixed continuously compounded interest rate r. We make an assumption on the stock price:

$$S_n = S_{n-1} + S_{n-1}\xi_n, \tag{Ia}$$

where ξ_n is a random variable which can take one of two values a_n and b_n satisfying the condition

$$-1 < a_n < b_n. \tag{Ib}$$

The condition (Ib) guarantees the positiveness of the price[1]: if $S_0 > 0$ then $S_n > 0$ for any n (we leave this simple exercise for you to check). Further, $a_n < b_n$ ensures that two corresponding realisations of the random variable ξ_n are different and hence, given a value of S_{n-1}, the price S_n can take two *different* values, otherwise (if $a_n = b_n$) S_n is deterministic. The values a_n and b_n may depend on S_{n-1} but not on S_n.

---

[1]As before, we follow the requirement that prices should be positive.

The **binary tree** (Ia) is illustrated in Fig. 7.1, where $0 < p_n < 1$ is the ('original' or 'actual' or 'market') probability that ξ_n takes the value a_n. We emphasise that both states of ξ_n have positive probabilities.

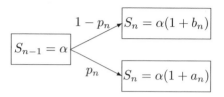

Fig. 7.1 Binary tree.

There is a special form of binary tree called a *recombinant tree* or *binomial tree*. In this case $a_n \equiv a$ and $b_n \equiv b$ for all n. In such a tree the same stock price can be attained in many different ways. This was introduced by Cox, Ross and Rubinstein (1979) and is known as the CRR model. Computationally, it is easier to work with than the general binary model and is quite adequate. The binary (and especially binomial) model is close, in a sense, to that of the Bernoulli scheme in classical Probability. Due to its simplicity, we can demonstrate ideas of option pricing without involving mathematics beyond secondary school.

Further, we make another natural assumption:

$$1 + a_n < e^{r\Delta t} < 1 + b_n, \qquad \text{(Ic)}$$

which is natural because it follows from the no-arbitrage condition. Indeed, suppose, for example, that

$$e^{r\Delta t} \leq 1 + a_n < 1 + b_n,$$

then

$$S_{n-1}e^{r\Delta t} \leq S_n,$$

from which $S_0 e^{r\Delta t} \leq S_1$, i.e. we have comparatively cheap borrowing in comparison with *a priori* knowledge of behaviour of the stock. Thus, if we buy an amount of stock at $t = 0$ for the price S_0 by borrowing money, we can sell the stock at the next discrete-time moment at a higher or equal price (taking into account the time value of money) with certainty. That is, we can make a risk-free profit. As before, we eliminate this possibility.

Exercise 7.1. By constructing an appropriate strategy, show that there is an arbitrage opportunity when $1 + a_n < 1 + b_n \leq e^{r\Delta t}$.

In what follows we work under the following set of assumptions in addition to I(a)–I(c) (they are analogous to the assumptions we used for pricing forwards in Section 3.2).

Further Assumptions (FA):

- The stock can be bought and sold in unlimited amounts for the existing price.
- Money can be borrowed and lent at the continuously compounded interest rate r in arbitrary size.
- There is no charge for holding positive or negative amounts of stock.
- There are no taxes or transaction costs.

The first two assumptions can be interpreted to be what we consider small investors.

7.1   The Mathematical Problems for European Options

Let $f(s)$ be a payoff function of an option (e.g., for a call $f(s) = (s - K)_+$).

1. What is the fair price one should pay at time $t = 0$ for an option (e.g., in the case of a European call: a premium to have a right to buy one unit of stock at cost K at the maturity time $t = T$)?
   This is the problem of *pricing the option*. We will give a rigorous definition of the fair price later, in Section 10.3. At this stage, we will understand under the fair price of an option the following: both writer and holder of the option should agree on this price, i.e. the price should allow the writer to meet his obligation according to the option (otherwise, he has a risk of losing money) and the price should correspond to the no-arbitrage principle.
2. How should the writer, who earns the premium initially, generate an amount $f(S_N)$ required to meet his obligation at time T?
   This is the problem of *hedging the option*. As we will see, these two problems are interlinked.

7.2   The One-Step (Single-Period) Binomial Model

As usual, we start consideration with the simplest case and price options within the single-period binomial model (I) (i.e. for $N = 1$, $T = \Delta t$).

Example 7.1. The current price of a certain stock is $S_0 = \$15$. A European call maturing in one month (i.e. $T = 1/12$) has strike price $K = \$18$. An investor believes that with probability of $1/2$ the stock price will be either $\$24$ or $\$12$. The riskless borrowing rate is zero, i.e. $r = 0$. By solving the equations $12 = 15(1 + a)$ and $24 = 15(1 + b)$, we find $a = -0.2$ and $b = 0.6$ in the equation (Ia). These data satisfy[2] the conditions (Ib) and (Ic).

The investor follows the logic of the law of large numbers from Section 2.1 and calculates the expected value of the option at the maturity time:

$$EC_1 = E(S_1 - K)_+ = \frac{1}{2}(24 - 18)_+ + \frac{1}{2}(12 - 18)_+$$

$$= \frac{1}{2} \times 6 + \frac{1}{2} \times 0 = 3 \ (\$).$$

So, the investor agrees to pay $\$3$ for the option, but is this a fair price? Due to what we have learned in Part I, we can guess that the answer is 'no'. The writer of the call the investor is buying could use the strategy (one of many) given in Table 7.1.

Table 7.1   The writer's strategy in Example 7.1.

| | $t = 0$ | $t = T$ | |
		$S_T = 12$	$S_T = 24$
Sell a call	3	0	−6
Buy stock	−15	12	24
Borrow	12	−12	−12
Total	0	0	6

Then the writer has a positive chance of making a profit with no risk of making a loss, i.e. there is arbitrage. The price of the option, $\$3$, is too high though it follows from the law of large numbers.

How do we price it correctly, i.e. in an acceptable way for both writer and holder of the option? Assume the framework of the model (I) with $N = 1$. Consider the writer's portfolio (ϕ, ψ), namely ϕ amount of stock S and ψ amount of money on the bank account B. If we buy the portfolio at time zero, it costs

$$V_0 = \phi S_0 + \psi. \tag{7.1}$$

At the next discrete-time moment, $t = \Delta t$, it would be worth one of two possible values as demonstrated in Fig. 7.2.

---

[2]Please check that this is true.

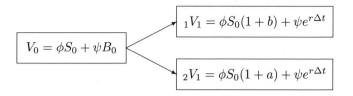

Fig. 7.2 Evolution of the price of the portfolio on one step of the binary tree.

Because of our aim to find a fair price, for every possible outcome, the writer must have

$$V_1 = f(S_1), \tag{7.2}$$

where $f(s)$ is the payoff function of the option considered (e.g., $f(s) = (s - K)_+$ for a call and $f(s) = (K - s)_+$ for a put). If at least one realisation is such that $V_1 < f(S_1)$, the writer will be unable to meet his obligation. If, however, $V_1 > f(S_1)$, he has a riskless profit with a positive probability.

We have two free variables ϕ and ψ and two equations, (7.1) and (7.2), which give us:

$$\phi S_0(1 + a) + \psi e^{r\Delta t} = f(S_0(1 + a)) := f_a$$
$$\phi S_0(1 + b) + \psi e^{r\Delta t} = f(S_0(1 + b)) := f_b,$$

and, solving them, we obtain

$$\phi = \frac{f_b - f_a}{S_0(b - a)}, \quad \psi = e^{-r\Delta t}[f_b - \frac{f_b - f_a}{b - a}(1 + b)]. \tag{7.3}$$

If the writer bought this (hedging) portfolio (ϕ, ψ) at $t = 0$ and held it, the equations (7.3) guarantee that he achieves his goal: if the stock moves up then the portfolio becomes worth f_b and if the stock moves down the portfolio becomes worth f_a.

Thus, the fair price (premium) the writer should be paid for that option is equal to

$$x = V_0 = \phi S_0 + \psi, \tag{7.4}$$

where ϕ and ψ are from (7.3). Having this amount of money the writer can meet his obligation to the holder if he follows the right strategy. At the same time, charging this premium, he cannot make a risk-free profit. We observe that using the portfolio (ϕ, ψ) we have *synthesised*[3] (replicated) the

---

[3] Recall that in Problems 15, 16 and 33 of Chapter 6 we have already encountered the idea of synthesising derivatives.

derivative, i.e. we created the payoff of the considered option using other financial instruments (the bank account and stock in this case).

In the above consideration we have not directly used the precise form of the payoff function. So, any European-type derivative with a payoff function $f(s)$[4], i.e. an option which promises to pay $f(S_T)$ at the maturity time T to its holder with S_T being the price of the underlying at T, can be constructed from an appropriate portfolio of a bank account and stock in advance.

We can now summarise these results as a theorem.

Theorem 7.1. *Consider the one-step $(N = 1)$ model (I) and suppose the assumptions (FA) hold. Then:*

1. At time zero the market price of an option with payoff $f(S_T)$ and maturity $T = \Delta t$ is

$$x = V_0 = \phi S_0 + \psi, \tag{7.5}$$

where

$$\phi = \frac{f_b - f_a}{S_0(b - a)}, \qquad \psi = e^{-r\Delta t}\left[f_b - \frac{f_b - f_a}{b - a}(1 + b)\right]. \tag{7.6}$$

2. The writer of the option can construct a portfolio whose value at time $T = \Delta t$ is exactly $f(S_1)$ by using the money received from the option to buy ϕ units of stock and deposit ψ on the bank account with (ϕ,ψ) from the equations (7.6).

Remark 7.1. Though our model allows randomness, arbitrage again dictates the rules of the game and leads to a deterministic answer not based on the strong law of large numbers, which is not helpful for discovering prices of financial products (like it was in Example 2.1 about a bookmaker). Note (see also Remark 3.1) that in finding the option price (7.5) we have used two instruments: bank account and stock which allowed us to discover the arbitrage price.

Remark 7.2. We did not use the probabilistic structure (the probability p of the stock price going down) except that both up and down movements can happen with positive probabilities – 'actual' (market) probabilities do not matter in our risk-free strategy.

---

[4]As an example additional to the considered plain vanilla call and put, a binary asset-or-nothing call option has the payoff $f(s) = \begin{cases} 0, & \text{if } s < K, \\ s, & \text{if } s \geq K. \end{cases}$

Example 7.2. *Continuation of Example 7.1.* What would be the fair price to pay for that call instead of $3? We have $S_0 = \$15$, $K = \$18$, $a = -0.2$ and $b = 0.6$. We also find that

$$f_b = (24 - 18)_+ = \$6, \quad f_a = (12 - 18)_+ = \$0,$$

then

$$\phi = \frac{6 - 0}{15(0.6 + 0.2)} = \frac{1}{2}, \quad \psi = 1 \times [6 - \frac{6}{0.6 + 0.2} \times 1.6] = -6$$

and

$$C_0 = V_0 = \frac{1}{2}S_0 - 6 = 7.5 - 6 = 1.5 \ (\$).$$

The arbitrage price of the call is $1.5. For any price higher than that the writer can make a risk-free profit. For any lower price, there will also be an arbitrage opportunity and the writer will not meet his obligation.

The writer's hedging strategy at time $t = 0$:

- borrow $6;
- use this money and the premium $1.5 to buy 1/2 stock unit.

Then, at the maturity time the writer breaks even independently of market movement. The strategy we used is self-financing.

Definition 7.1. A ***self-financing strategy*** *is a strategy which never needs to be topped up with extra cash nor can ever afford withdrawals.*

We will put observations made in this chapter on a more rigorous footing after we revise and learn some elements of Probability theory.

Chapter 8

Elements of Probability Theory

Some notions from Probability theory have already been used but now we need more formal and rigorous definitions. We start with a probabilistic model (or probability space) in its simplest form.

8.1 Finite Probability Spaces or Probabilistic Models with Finite Numbers of Outcomes

A finite probability space is used for modelling a random experiment with finitely many possible outcomes.

Example 8.1. In the context of the binary model (I) we considered, we essentially just tossed a coin at each step and for a fixed number of steps N, we tossed the coin N times. If, e.g., we toss the coin three times then the set of all possible outcomes is

$$\Omega = \{'HHH','HHT','HTH','HTT','THH','THT','TTH','TTT'\},$$

where $'T'$ is for tails and $'H'$ is for heads.

In general, consider an experiment where all of the possible results are included in a finite number of outcomes $\omega_1, \ldots, \omega_M$.

Definition 8.1. *Outcomes of an experiment* $\omega_1, \ldots, \omega_M$ *are called **elementary events** or **sample points** and the finite set*

$$\Omega = \{\omega_1, \ldots, \omega_M\}$$

*is called the **space of elementary events** or the **sample space**.*

Choosing an appropriate sample space is the first step in formulating a probabilistic model of an experiment. In practice we are often interested not

75

in what particular outcome occurs as the result of a trial but whether the outcome belongs to some subset of the set of all possible outcomes.

Example 8.2. (*Continuation of Example 8.1*). We play a game, where I win if at least two heads appear, i.e. I win in the case of the following event:

$$A = \{'HHH','HHT','HTH','THH'\}.$$

Definition 8.2. *Events* are all subsets $A \subset \Omega$, for which under the condition of the experiment, one can conclude that either 'the outcome $\omega \in A$' or 'the outcome $\omega \notin A$'.

Starting from a given collection of sets that are events, one can form new events using the logical connectives from Table 8.1.

Table 8.1   The logic connectives and set-theoretical operations.

Logic	Set theory	Notation
Or	Union	$A \cup B$
And	Intersection	$A \cap B$
Not	Complement	$\bar{A} = \Omega \setminus A$

The sets A and $\bar{A}$ have no points in common and hence $A \cap \bar{A} = \emptyset$ is empty.

In Probability theory, $\emptyset$ is called an **impossible event** and the set Ω is called the **certain event** or **sure event**.

Consider a collection $\mathcal{A}_0$ of sets $A \subseteq \Omega$. Using the set-theoretical operations from Table 8.1, one can form a new collection of sets (which are events) from the elements of $\mathcal{A}_0$. These sets are again events. If we adjoin the certain event Ω and the impossible event $\emptyset$, we obtain a collection of sets $\mathcal{A}$ which is an algebra.

Definition 8.3. *An **algebra** is a collection of subsets of Ω such that*
1. $\Omega \in \mathcal{A}$;
2. *if $A \in \mathcal{A}$ and $B \in \mathcal{A}$ then the sets $A \cup B$, $A \cap B$ and $A \setminus B$ also belong to $\mathcal{A}$.*

Example 8.3. Let us look at three examples of algebras which we will use in the future.

1. $\mathcal{A} = \{\Omega, \emptyset\}$ – trivial algebra.

2. $\mathcal{A} = \{A, \bar{A}, \Omega, \emptyset\}$ – the algebra generated by A.

3. $\mathcal{A} = \{A : A \subseteq \Omega\}$ – the collection of all the subsets of Ω (including $\emptyset$).

Exercise 8.1. Check that all three $\mathcal{A}$ in the above example satisfy the conditions from Definition 8.3.

We will mostly be working with an algebra as described in point 3 of the above example.

We have introduced two elements of the probability space $(\Omega, \mathcal{A})$. We need the third one. We assign a **weight** to each sample point $\omega_i \in \Omega$, $i = 1, \ldots, M$. This weight is denoted by $p(\omega_i)$ and called the **probability of the outcome** ω_i or the **elementary probability**. We assume that the two **axioms** hold:

1. $0 \le p(\omega_i)$ for all $\omega_i \in \Omega$ (non-negativity).
2. $\sum_{i=1}^{M} p(\omega_i) = 1$ (normalisation).

Example 8.4. (*Continuation of Examples 8.1 and 8.2*). We may assume that our experiments with the coin were organised in such a way that all the outcomes are equally probable. Then

$$p(\omega_i) = \frac{1}{8}, \quad i = 1, \ldots, 8.$$

Starting from the elementary probabilities $p(\omega_i)$, we define the **probability** $P(A)$ of any event $A \in \mathcal{A}$ by

$$P(A) = \sum_{\{i : \omega_i \in A\}} p(\omega_i).$$

Example 8.5. (*Continuation of Examples 8.1, 8.2 and 8.4*). Compute $P(A)$ with A from Example 8.2 and the elementary probabilities from Example 8.4:

$$P(A) = \sum_{\{i : \omega_i \in A\}} p(\omega_i) = \sum_{i=1}^{4} \frac{1}{8} = \frac{1}{2}.$$

Finally, we arrive at the definition of a probabilistic model.

Definition 8.4. *Let* $\Omega = \{\omega_i, \ldots, \omega_M\}$ *with a finite* M *be the sample space,* $\mathcal{A}$ *be an algebra of subsets of* Ω *and* $P = \{P(A); A \in \mathcal{A}\}$ *be probability of events from* $\mathcal{A}$. *Then the triple* $(\Omega, \mathcal{A}, P)$ *defines a* **probabilistic model** *or* **probability space** *of experiments with space* Ω *of outcomes and algebra* $\mathcal{A}$ *of events.*

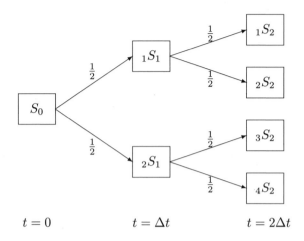

$$t = 0 \qquad\qquad t = \Delta t \qquad\qquad t = 2\Delta t$$

Fig. 8.1   The two-period binomial model from Example 8.6.

In what follows, we always assume that we have an appropriate probability space $(\Omega, \mathcal{A}, P)$. Now we introduce (or revise) some important notions of Probability theory.

Definition 8.5.[1]

1. *Events A and B are called **independent** if $P(AB) = P(A)P(B)$.*
2. *Events $A_1, A_2, \ldots, A_n$ are called **independent** if for any subset of different indices $i_1, i_2, \ldots, i_k$, $k \geq 1$,*

$$P(A_i \cdots A_{i_k}) = \prod_{j=1}^{k} P(A_{i_j}).$$

3. *If $P(B) > 0$ then the **conditional probability** $P(A|B)$ is defined as*

$$P(A|B) = \frac{P(AB)}{P(B)}.$$

Example 8.6. Consider a two-period binomial model given in Fig. 8.1. Suppose that the movements of the price are independent (we toss a coin

---

[1]Here let us quote A. N. Shiryaev (see Shiryaev, 1996, p. 27): 'In a sense, the concept of independence... plays a central role in Probability theory: it is precisely this concept that distinguishes Probability theory from the general theory of measure spaces.'

twice independently). Assume that each movement 'up' or 'down' happens with the probability of $1/2$. Find the probability that two 'up' movements of the price happen, assuming that

1. the first movement is 'up';
2. there is at least one 'up' movement.

Answer. The sample space is $\Omega = \{'DD','DU','UD','UU'\}$ and
$$P('DD') = P('DU') = P('UD') = P('UU') = \frac{1}{4}.$$

1. Let A be the event that the first movement is 'up': $A = \{'UD','UU'\}$. We have
$$P('UU'|A) = \frac{P('UU')}{P(A)} = \frac{P('UU')}{P('UD') + P('UU')} = \frac{\frac{1}{4}}{\frac{1}{4} + \frac{1}{4}} = \frac{1}{2}.$$

2. *Exercise.* Please do it yourself. For your verification the answer is $1/3$.

8.2 Random Variables: Definition and Expectation

Definition 8.6. *A real-valued* **random variable** $\xi = \xi(\omega)$ *is a real-valued function on Ω that is $\mathcal{A}$-measurable.*

$\mathcal{A}$**-measurable** means that for any $x \in \mathbf{R}$
$$\{\omega \in \Omega : \xi(\omega) \leq x\} \in \mathcal{A}.$$
The importance of this condition is that if a probability measure P is defined on $(\Omega, \mathcal{A})$, it then makes sense to speak of the probability of the event $\{\xi \leq x\}$ since we know how to assign probabilities to events from $\mathcal{A}$.

Example 8.7. The function
$$I(A) \equiv I_A(\omega) = \begin{cases} 1, & \omega \in A \\ 0, & \omega \notin A, \end{cases} \quad \text{for some } A \in \mathcal{A},$$
is called the **indicator** or **characteristic function** of a set $A \in \mathcal{A}$. Then $\xi := I_A(\omega)$ is an example of a random variable. In Example 8.2, $\xi = I_A(\omega)$ gives 1 for win and 0 for loss.

Let $\xi(\omega)$ be a random variable with values in the set $X = \{x_1, \ldots, x_k\}$ defined on a finite probability space. If we put $A_i = \{\omega \in \Omega : \xi(\omega) = x_i\}$, we can represent this **discrete** random variable as
$$\xi(\omega) = \sum_{i=1}^{k} x_i I_{A_i}(\omega),$$

where the sets $A_1, \ldots, A_k$ form a **decomposition** of Ω (i.e. they are pairwise disjoint[2] and their sum is Ω).

Let us now consider some characteristics of random variables.

Definition 8.7. *The function*

$$F_\xi(x) := P(\{\omega : \xi \leq x\}), \quad x \in \mathbf{R},$$

*is called the **distribution function** of ξ.*

For a discrete random variable ξ with values in the set $X = \{x_1, \ldots, x_k\}$:

$$F_\xi(x) = \sum_{\ell : x_\ell \leq x} P(\{\omega : \xi = x_\ell\}) \equiv \sum_{\ell : x_\ell \leq x} p_\ell,$$

*where $p_\ell \equiv P_\xi(x_\ell) := P(\{\omega : \xi = x_\ell\})$, $\ell = 1, \ldots, k$, is called the **probability distribution** of a discrete random variable ξ.*

Example 8.8. (*Bernoulli distribution*). Consider a random variable ξ that takes values from $X = \{a, b\}$, $a, b \in \mathbf{R}$, and

$$p_1 = P(\{\omega : \xi = a\}) = p, \quad p_2 = P(\{\omega : \xi = b\}) = 1 - p,$$

where $p \in [0, 1]$. We call such ξ a Bernoulli random variable (e.g., ξ_n in the equation (Ia) are Bernoulli random variables).

Definition 8.8. *The **expectation operator** $E_P(\cdot)$ has one parameter, a measure P, and in the discrete case*

$$E_P \xi(\omega) := \sum_{i=1}^{M} \xi(\omega_i) p(\omega_i).$$

Properties of expectations[3]

Let ξ and η be random variables and a and b be some real numbers. The following properties hold:

1. $Ea = a$.
2. $E(a\xi + b\eta) = aE\xi + bE\eta$ (linearity).
3. If $\xi \leq \eta$, then $E\xi \leq E\eta$.

---

[2]A collection of sets is pairwise disjoint if every set is disjoint from every other set in the collection.

[3]See their proofs in, e.g., Shiryaev (1996).

Example 8.9. (*Can we link option pricing with Probability language?*). Let us return to the single-period binomial model and pricing European options. We have the probability space $(\Omega, \mathcal{A}, P)$, where

$$\Omega = \{'D','U'\}, \quad P('D') = p, \quad P('U') = 1 - p.$$

The expectation of the payoff at the maturity time T with respect to the 'original' probability is

$$E_P f(S_T) = p f_a + (1 - p) f_b. \tag{8.1}$$

We know that this is not a very useful characteristic, therefore, we go back to Theorem 7.1 and manipulate the formulae in order to see whether we can write the option price as an expectation.

We have (see Theorem 7.1):

$$V_0 = \phi S_0 + \psi = \frac{f_b - f_a}{b - a} + e^{-r\Delta t} \left[f_b - \frac{f_b - f_a}{b - a}(1 + b) \right]. \tag{8.2}$$

Let us compare the equations (8.1) and (8.2) and look at the weights of f_a and f_b in (8.2):

$$V_0 = e^{-r\Delta t} \left[f_b \frac{e^{r\Delta t} - (1 + a)}{b - a} + f_a \frac{(1 + b) - e^{r\Delta t}}{b - a} \right].$$

Introduce

$$q := \frac{e^{r\Delta t} - (1 + a)}{b - a}.$$

Due to the assumption (Ic), we have $0 < q < 1$ (*Exercise*: please check this yourself). Further,

$$\frac{(1 + b) - e^{r\Delta t}}{b - a} = 1 - \frac{e^{r\Delta t} - (1 + a)}{b - a} = 1 - q \quad \text{and } 0 < 1 - q < 1.$$

Thus,

$$x = V_0 = e^{-r\Delta t} \left[f_b q + f_a(1 - q) \right],$$

where q, $1 - q$ satisfy the probability axioms. We can introduce the corresponding new probability space $(\Omega, \mathcal{A}, Q)$, where Ω and $\mathcal{A}$ are the same as before but Q is the new measure. Then,

$$x = e^{-r\Delta t} E_Q f(S_T).$$

This is the expectation discounted by the growth of money on the bank account. But this expectation is not due to the 'original' probabilities p and

$1 - p$, it has (almost[4]) no connection with the 'original' (market) probability. The expectation is with respect to the new probability:

$$Q(\xi = b) = q, \quad Q(\xi = a) = 1 - q.$$

This measure Q is called **risk-neutral, risk-free, risk-adjusted** or im-**plied**. Correspondingly, the probabilities q and $1 - q$ are called **implied probabilities** (see also Example 2.1 in Section 2.1). Now we can reformulate Theorem 7.1.

Theorem 8.1. *Consider the one-step ($N = 1$) of the binary model (I) and suppose the assumptions (FA) hold. Then:*

1. *At time zero the market price of an option with payoff $f(s)$ and maturity $T = \Delta t$ is*

$$x = V_0 = e^{-rT} E_Q f(S_T),$$

where the risk-neutral probability Q is defined by

$$Q(\xi = b) = q = \frac{e^{r\Delta t} - (1 + a)}{b - a}, \quad Q(\xi = a) = 1 - q. \qquad (8.3)$$

2. *As in Theorem 7.1.*

Remark 8.1. In the discrete-time case, which we are considering, the introduction of the probability measure Q can be avoided as the results can be obtained using elementary Algebra (we got the price and hedging strategy without Q in Section 7.2). However, in the continuous-time case, where the Algebra does not work, a new measure like Q is essential.

Remark 8.2. We have separated the two components that would normally (e.g., in applications from Physics and Biology) be seen as interconnected parts of the indivisible whole – the probability of a move and where the move goes to. However the no-arbitrage principle breaks this natural union. In pricing an option we were interested only in what might happen and not in the actual probabilities of the outcomes (cf. Example 2.1 about the bookmaker). We did not need the 'original' (market) measure P in order to find the measure Q, which allowed the risk-free construction. That measure Q is a function of S and (essentially) not a function of P. This separation of process and measure is not artificial – it is fundamental to everything we do in Financial Mathematics. In Section 2.1 the strong law of large numbers failed in managing the risk because it used the union of outcomes and their probabilities, not possible outcomes alone.

---

[4]Except that we should require that both movements of the price (up and down) are possible, i.e. that $p > 0$ and $1 - p > 0$. See further details at the beginning of Section 9.3.

Example 8.10.

1. We have for ξ_1 from (I) and Q from (8.3):

$$E_Q \xi_1 = a(1 - q) + bq = e^{r\Delta t} - 1.$$

We see that this expectation depends only on the growth of money with time.

2. Compute $E_Q S_1$ for S_1 from (I):

$$E_Q S_1 = E_Q[S_0(1 + \xi_1)] = S_0 E_Q(1 + \xi_1) = e^{r\Delta t} S_0.$$

Hence

$$S_0 = E_Q e^{-r\Delta t} S_1.$$

We see that the expectation of the stock price under the risk-neutral probability discounted by the growth of money with time equals the stock price at the previous time moment. We will use this observation in future.

8.3 Random Variables: Independence and Conditional Expectation

We continue to build/revise our knowledge of Probability theory. Consider two discrete random variables ξ and η with values from $X = \{x_1, \ldots, x_m\}$ and $Y = \{y_1, \ldots, y_\ell\}$, respectively.

Definition 8.9. *The set of probabilities*

$$P_{\xi\eta}(x, y) = P(\xi = x, \eta = y),$$

*where $x \in X$ and $y \in Y$, is called the **joint probability distribution** of ξ and η.*

Definition 8.10. *The random variables ξ and η are said to be **independent** if*

$$P(\{\omega : \xi = x, \eta = y\}) = P(\{\omega : \xi = x\})P(\{\omega : \eta = y\}) \quad x \in X, y \in Y,$$

(or, equivalently, $P_{\xi\eta}(x, y) = P_\xi(x)P_\eta(y)$).

Property 8.1. *If ξ and η are independent then*

$$E_P(\xi\eta) = E_P\xi E_P\eta.$$

Exercise 8.2. Prove this property (see also Shiryaev, 1996).

Exercise 8.3. Give an example of ξ and η such that they are dependent but $E_P(\xi\eta) = E_P\xi E_P\eta$.

Definition 8.11. *(i) The quantity*

$$E_P(\xi|\eta = y) := \sum_{j=1}^{m} x_j \frac{P_{\xi\eta}(x_j, y)}{P_\eta(y)} = \sum_{j=1}^{m} x_j P(\xi = x_j|\eta = y), \ x_j \in X, \ y \in Y,$$

*is called the **conditional expectation** of ξ with respect to $\{\eta = y\}$.*
*(ii) The random variable $E_P(\xi|\eta)$ is called the **conditional expectation** of ξ with respect to η if $E_P(\xi|\eta)$ is equal to $E_P(\xi|\eta = y)$ on every set $\{\omega : \eta = y\}$, $y \in Y$; or, in other words,*

$$E_P(\xi|\eta)(\omega) := \sum_{i=1}^{\ell} \sum_{j=1}^{m} x_j P(\xi = x_j|\eta = y_i) \times I_{\{\omega:\eta=y_i\}}(\omega).$$

We have extended the notion of the expectation operator from Definition 8.8, which has one parameter (measure), to the conditional expectation operator which has two parameters. The conditional expectation $E_P(\xi|\eta)$ depends not only on measure P but also on the random variable η. We emphasise that while $E_P(\xi|\eta = y)$ is a number, $E_P(\xi|\eta)$ is a random variable.

Example 8.11. Consider the two-step binomial model (the tree diagram is plotted in Fig. 8.2):

$$S_n = S_{n-1}(1 + \xi_n), \ n = 1, 2,$$

where ξ_1 and ξ_2 are independent random variables taking the values $a_n \equiv a$, $b_n \equiv b$, $n = 1, 2$ with probability $P(\xi_n = a) = p$, $P(\xi_n = b) = 1 - p$, respectively. It follows that $\dfrac{S_1}{S_0}$ and $\dfrac{S_2}{S_1}$ are independent (note that S_2 and S_1 are dependent).

The following notation is used here and in what follows: S_1 is a random variable taking values in $\{ {}_1S_1, {}_2S_1\}$, S_2 is a random variable taking values in $\{ {}_1S_2, {}_2S_2, {}_3S_2, {}_4S_2\}$, and so on, S_n is a random variable taking values in $\{ {}_1S_n, {}_2S_n, \ldots, {}_{2^n}S_n\}$.

Calculate the conditional expectations:

- $E_P(S_2|S_1 = {}_1S_1)$.
- $E_P(S_2|S_1)$.

Answer. Observe that the random variable S_1 takes two values:

$${}_1S_1 = S_0(1 + b) \text{ and } {}_2S_1 = S_0(1 + a).$$

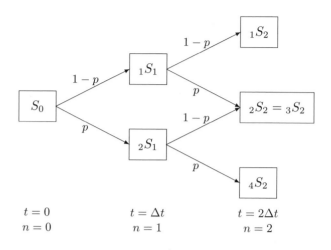

$$t = 0 \qquad t = \Delta t \qquad t = 2\Delta t$$
$$n = 0 \qquad n = 1 \qquad n = 2$$

Fig. 8.2 The tree diagram for Example 8.11.

We need to compute $E_P(S_2|S_1 = {_i}S_1)$ with $i = 1, 2$. To this end, we first compute the expectation of S_2 assuming that at the time $n = 1$ the price was ${_1}S_1$. In this case S_2 can take values ${_1}S_2$ with probability $1 - p$ and ${_2}S_2$ with probability p, then

$$E_P(S_2|S_1 = {_1}S_1) = (1 - p)\ {_1}S_2 + p\ {_2}S_2 = {_1}S_1(1 + b + p(a - b)).$$

Analogously,

$$E_P(S_2|S_1 = {_2}S_1) = {_2}S_1(1 + b + p(a - b)).$$

Thus

$$E_P(S_2|S_1) = S_1(1 + b + p(a - b)).$$

Notice that $E_P(S_2|S_1)$ is a random variable while $E_P(S_2|S_1 = {_1}S_1)$ is a number.

Let us give an additional interpretation of expectation and conditional expectation, which can also be used as their alternative definitions. Consider a random variable ξ for which the second moment is finite: $E\xi^2 < \infty$[5].

---

[5]Though for discrete random variables considered in this chapter this condition holds automatically in all situations of our interest, we state it for completeness of the exposition.

Suppose one would like to estimate the random variable ξ (which we can view as a measurement in an experiment) by a single constant c. Then $E(\xi - c)^2$ might be used as a measure of how good the estimate is, with small numbers indicating a better estimate.

Example 8.12. Find c which gives the best estimate of ξ in the mean-square sense, i.e. the c which attains the least mean-square error $E(\xi - c)^2$ and hence is the best mean-square estimate among constants.

Answer. Denote $m = E\xi$. We have

$$
\begin{aligned}
E(\xi - c)^2 &= E((\xi - m) + (m - c))^2 \qquad\qquad (8.4) \\
&= E(\xi - m)^2 + E[2(\xi - m)(m - c)] + E(m - c)^2 \\
&= E(\xi - m)^2 + E(m - c)^2.
\end{aligned}
$$

Hence, for a random variable ξ with $E\xi^2 < \infty$, $m = E\xi$ is the unique constant which attains the least mean-square error. Note that this example also gives us the interpretation of variance as the mean-square error of estimation of a random variable by its mean.

Now we interpret conditional expectation. Suppose there are two random variables ξ and η of our interest coming out of a particular experiment and we want to know the value of ξ but only have the value of η (e.g., you can view η to be a price of a commodity today and ξ as its price tomorrow). What is the best way to use this information η for estimating the unobservable ξ? An estimate of ξ given η is a random variable $f(\eta)$. To say which $f(\eta)$ is the best, we need a criterion. We will again measure the error in the mean-square sense. We say that an estimate g of ξ is optimal in the mean-square sense given η if

$$
E(\xi - g(\eta))^2 \leq E(\xi - f(\eta))^2
$$

for all possible estimates $f(\eta)$ of ξ. Let us note that due to Definition 8.11 the conditional expectation $E(\xi|\eta)$ is an estimate of ξ given η. Moreover, one can prove (see, e.g., Fristedt *et al.*, 2007, pp. 29–30) that the conditional expectation $E(\xi|\eta)$ is the mean-square *optimal* estimate of ξ obtained via observations of η. On the one hand, this gives an interpretation of conditional expectations, on the other hand we can use this interpretation as an alternative definition.

8.4 Properties of Conditional Expectations

It can be proved (see, e.g., Shiryaev, 1996; Williams, 2001) that conditional expectations have the following properties. Assume that random variables ξ, η and ζ have the finite means: $E|\xi|$, $E|\eta|$, $E|\zeta| < \infty$, and a, $b \in \mathbf{R}$ are some constants. Then

1. $E\xi = E(E(\xi|\eta))$.
2. If ξ and η are independent,

$$E(\xi|\eta) = E\xi.$$

3. Since by definition $E(\xi|\eta)$ is a function of η, the conditional expectation can be interpreted as a prediction of ξ given the information from the 'observed' random variable η.
4. $E(\xi|\xi) = \xi$ and $E(\xi\eta|\xi) = \xi E(\eta|\xi)$.
5. $E(a\xi + b\zeta|\eta) = aE(\xi|\eta) + bE(\zeta|\eta)$ (linearity).
6. $E(a|\eta) = a$.
7. If $\xi \leq \zeta$, $E(\xi|\eta) \leq E(\zeta|\eta)$.
8. $|E(\xi|\eta)| \leq E(|\xi|\,|\eta)$.

We note that the interpretation of conditional expectations given at the end of the previous section makes many of these properties natural. For instance, let us look at the second property. If ξ and η are independent, then knowing the value of η does not add any information about ξ and its best mean-square estimate given η is just its best mean-square approximation by a constant, which is $E\xi$.

Example 8.13. (*Continuation of Example 8.11*). Using the properties of conditional expectations, we get

$$E_P(S_2|S_1) = E_P(S_1(1 + \xi_2)|S_1) \stackrel{\text{property } 4}{=} S_1 E_P(1 + \xi_2|S_1)$$
$$\stackrel{\text{property } 2}{=} S_1 E_P(1 + \xi_2) = S_1(1 + b + p(a - b)).$$

In the next definition we generalise the definitions of conditional expectations from the previous section to the case when we are interested in knowing the value of ξ given value of a random *vector* η (e.g., you can view η to be prices of a commodity at the end of the trading days during the last week and ξ as its price tomorrow).

Definition 8.12. *A **conditional expectation** of a discrete random variable ξ, given a random vector $\eta = (\eta_1, \ldots, \eta_n)$ with values from $Y =$*

$\{y_1, \ldots, y_l\}$ *and which components* η_i *are discrete random variables, is*

$$E_P(\xi|\eta = y) = \sum_{j=1}^{m} x_j P(\xi = x_j|\eta = y),\, y \in Y,$$

and

$$E_P(\xi|\eta) = E_P(\xi|\eta_1, \ldots, \eta_n) := \sum_{j=1}^{m} \sum_{i=1}^{l} x_j P(\xi = x_j|\eta = y_i) I_{\{\omega:\eta=y_i\}},$$

where

$$P(\xi = x|\eta = y) = \frac{P(\xi = x, \eta = y)}{P(\eta = y)}.$$

Chapter 9

Discrete-Time Stochastic Processes

Consider random variables on a (finite) probability space $(\Omega, \mathcal{A}, P)$. In market models (as in any other dynamic models) we deal with families of random variables depending on a parameter (e.g., S_n stock price at $t = t_n$ has n as a parameter).

Definition 9.1. *Random function* $\zeta_t(\omega)$ *is a family of random variables, depending on the parameter* $t \in \mathbb{T}$. *When the set* $\mathbb{T}$ *is a subset of a real line* $\mathbf{R}$ *and the parameter t is interpreted as time, we say that* $\zeta_t(\omega)$ *is a* **random (stochastic) process** *(instead of a random function).*

If $\mathbb{T}$ consists of integers, one often says that $\zeta_t(\omega)$ is a *random sequence* or a *discrete-time stochastic process*. If we fix the elementary event in $\zeta_t(\omega)$ we obtain the non-random function $\zeta_t(\cdot)$ which is called the **trajectory** or **realisation** of the random process. Note that if we fix t then we obtain a random variable.

Since we have time, we have the history/information flow and we need to express this in mathematical terms. The term which is used for this purpose in Stochastic Analysis is **filtration** $\{\mathcal{A}_n\}$.

Definition 9.2. *(Discrete case).* **Filtration** *is an increasing sequence of algebras included in* $\mathcal{A}$: $\mathcal{A}_0, \mathcal{A}_1, \ldots, \mathcal{A}_N$:

$$\mathcal{A}_0 \subseteq \mathcal{A}_1 \subseteq \cdots \subseteq \mathcal{A}_N \subseteq \mathcal{A}. \tag{9.1}$$

An algebra $\mathcal{A}_n$, $n = 0, 1, \ldots, N$, can be seen as information available at time n. Filtration is sometimes called the algebra of events up to time n. In general, the information used by us to make a decision increases with time. Increase of the sets over time in (9.1) reflects the assumption that past data are never forgotten.

What is $\{\mathcal{A}_n\}$ *in our stock model* (I)? Recall that our market consists of a deterministic bank account and a stock whose price is modelled by a random process S_k, $k = 0, 1, 2, \ldots$. As the corresponding filtration to time n, we can take the set $\mathcal{A}_n \doteq \mathcal{A}_n^S$ of all the events which depend only on the values of the process S_k until this time, i.e. which depend on $(S_0, S_1, \ldots, S_n)$ only. These events are paths from 0-node to nodes at time $t = t_n$. In other words, $\mathcal{A}_n$ is the history of the stock until the time n of the tree or, we can say, $\mathcal{A}_n$ consists of all the available information on the market until time n. If $\mathcal{A}_n$ encodes precisely all of the information about the evolution of stochastic process S_k up until time n, then this information (or filtration) is called the **natural filtration** associated to the stochastic process $\{S_k\}_{k \geq 0}$.

Example 9.1. Let us take a two-step price model S_0, S_1, S_2 on a probability space $(\Omega, \mathcal{A}, P)$ with $\Omega = \{'UU', 'UD', 'DU', 'DD'\}$, where U and D mean movements 'up' and 'down' of the price at the corresponding time step and $\mathcal{A}$ is the collection of all subsets of Ω. Then, $\mathcal{A}_1$ is an algebra generated by the events

$$G_1 = \{\omega \in \Omega \colon S_1(\omega) = {}_1 S_1\} = \{'UU', 'UD'\} = 'U*',$$
$$G_2 = \{\omega \in \Omega \colon S_1(\omega) = {}_2 S_1\} = \{'DU', 'DD'\} = 'D*'.$$

Noting that $G_1 \cup G_2 = \Omega$, $G_1 \cap G_2 = \varnothing$, $\Omega \backslash G_1 = G_2$ and $\Omega \backslash G_2 = G_1$, we conclude that the algebra $\mathcal{A}_1 = \{\varnothing, \Omega, G_1, G_2\} \subset \mathcal{A}$.

In general, suppose that $\mathbf{S}_n := (S_0, S_1, \ldots, S_n)$ is a random vector which takes values ${}_i\mathbf{S}_n$, $i = 1, \ldots, 2^n$. Then, $\mathcal{A}_n$ is an algebra generated by the collection of events (or, in other words, the minimum algebra containing the collection of events) $G_i = \{\omega \in \Omega \colon \mathbf{S}_n(\omega) = {}_i\mathbf{S}_n\}$, $i = 1, \ldots, 2^n$.

Intuitively, if we say that $\mathcal{A}_n$ is known to us, this means that we know by which path we arrive at time n.

Definition 9.3. *The quadruple* $(\Omega, \mathcal{A}, \{\mathcal{A}_i\}_{i \geq 0}, P)$ *is called a **filtered probability space**.*

Definition 9.4. *We say that a stochastic process* ζ_n *is **adapted** to the filtration* $\{\mathcal{A}_i\}_{i \geq 0}$ *if* ζ_n *is* $\mathcal{A}_n$*-measurable for each* n.

Recall that ζ_n being $\mathcal{A}_n$-measurable means that the event $\{\omega \in \Omega : \zeta_n(\omega) \leq x\} \in \mathcal{A}_n$ for any $x \in \mathbf{R}$. We can also interpret Definition 9.4 in the following way: if the information contained in $\mathcal{A}_n$ is available to us, then we know the value of ζ_n. We will often use the notation $\zeta_n \in \mathcal{A}_n$ as a short writing for 'ζ_n being adapted to $\{\mathcal{A}_i\}_{i \geq 0}$'.

9.1 Conditional Expectation of a Random Variable Given Information

In Chapter 8 we considered the notions of expectations $E(\cdot) = E_P(\cdot)$ and the conditional expectations with respect to a random variable or a random vector $E(\cdot|\eta)$. Here, we introduce conditional expectation given information. Recall that we are dealing with finite probability spaces.

Definition 9.5. *Let $\mathcal{G} \subset \mathcal{A}$ be a minimal algebra containing a finite collection of disjoint events $G_1, G_2, \ldots, G_k$, the sum of which is equal to Ω and for all $i = 1, \ldots, k$ the probability $P(G_i) > 0$. A **conditional probability** of an event A with respect to $\mathcal{G}$ is a random variable $P(A|\mathcal{G})$ which is equal to $P(A|G_i)$ on the set G_i, i.e.*

$$P(A|\mathcal{G}) = P(A|\mathcal{G})(\omega) = \sum_{i=1}^{k} I_{G_i}(\omega)P(A|G_i). \qquad (9.2)$$

Recall that conditional probability $P(A|G_i)$ was defined in Definition 8.5. Consider a discrete random variable ξ with values from $X = \{x_1, \ldots, x_l\}$.

Definition 9.6. *A **conditional expectation** of ξ given algebra $\mathcal{G} \subset \mathcal{A}$ is*

$$E_P(\xi|\mathcal{G}) := \sum_{i=1}^{M} \xi(\omega_i)P(\omega_i|\mathcal{G}) = \sum_{j=1}^{l} x_j P(\xi = x_j|\mathcal{G}). \qquad (9.3)$$

Note that $P(A|\mathcal{G})$ and $E_P(\xi|\mathcal{G})$ are $\mathcal{G}$-measurable random variables. Also recall that M is the number of elementary events ω_i in the sample space Ω.

Analogously to the interpretation of conditional expectations with respect to a random variable given in Section 8.3, we can interpret conditional expectation given an algebra which is defined above. Let ξ be a random variable with $E|\xi|^2 < \infty$ and defined on $(\Omega, \mathcal{A}, P)$. A random variable $\theta := E_P(\xi|\mathcal{G})$ is such that $E_P|\xi - E_P(\xi|\mathcal{G})|^2 \leq E_P|\xi - \eta|^2$ for any η defined on $(\Omega, \mathcal{G}, P)$[1] and with the finite second moment $E|\eta|^2 < \infty$.

Remark 9.1. A general definition of conditional expectation with respect to an algebra is rather technical and it requires knowledge of the Lebesgue integral (see, e.g., Shiryaev, 1996; Etheridge, 2002; Klebaner, 2005; Krylov, 2002), which we do not consider in this course.

---

[1] By defined on $(\Omega, \mathcal{G}, P)$, we mean that events $\{\omega \in \Omega : \eta(\omega) \leq x\} \in \mathcal{G}$.

What is $E_P(\xi|\mathcal{A}_n)$ in our case? Let us exploit the notion of conditional expectation with respect to algebra for our purposes, namely to define a conditional expectation of ξ given information $\mathcal{A}_n$ associated with the model (I), in other words, with respect to the natural filtration $\mathcal{A}_n$ associated to the price process $\{S_k\}_{k \geq 0}$. As it was said before, this information $\mathcal{A}_n$ is determined by the random vector $\mathbf{S}_n := (S_0, S_1, \ldots, S_n)$. More precisely, if $_i\mathbf{S}_n$, $i = 1, \ldots, k$, are all possible paths from 0-node to nodes at time $t = t_n$, then one can take $\mathcal{A}_n$ as the minimal algebra containing the collection of events[2] $G_i = \{\omega \in \Omega : \mathbf{S}_n(\omega) = {}_i\mathbf{S}_n\}$, $i = 1, \ldots, k$. Hence

$$P(A|\mathcal{A}_n) \overset{\text{Definition 9.5}}{=} \sum_{i=1}^{k} I_{G_i}(\omega) P(A|G_i) \tag{9.4}$$

$$\overset{\text{definition of } G_i}{=} \sum_{i=1}^{k} I_{G_i}(\omega) P(A|\mathbf{S}_n = {}_i\mathbf{S}_n).$$

From here and Definitions 9.6 and 8.12, we obtain

$$E_P(\xi|\mathcal{A}_n) \overset{\text{Definition 9.6}}{=} \sum_{j=1}^{l} x_j P(\xi = x_j|\mathcal{A}_n) \tag{9.5}$$

$$\overset{(9.4)}{=} \sum_{j=1}^{l} \sum_{i=1}^{k} x_j I_{\{\omega : \mathbf{S}_n = {}_i\mathbf{S}_n\}}(\omega) P(\xi = x_j|\mathbf{S}_n = {}_i\mathbf{S}_n)$$

$$\overset{\text{Definition 8.12}}{=} E_P(\xi|S_0, S_1, \ldots, S_n).$$

Then, *in our context*, we can simplify Definition 9.6. In what follows, we use the below definition of a conditional expectation given information.

Definition 9.7. *(For our purposes!). A **conditional expectation** of ξ given information $\mathcal{A}_n$ is*

$$E_P(\xi|\mathcal{A}_n) := E_P(\xi|S_0, S_1, \ldots, S_n).$$

Note that by the agreement ξ is $\mathcal{A}$-measurable and the process S_k is $\{\mathcal{A}_i\}$-adapted. The quantity $E_P(f(S_m)|\mathcal{A}_n)$, $n \leq m$, is the expectation along the later portion of paths which have their initial segment determined by $\mathcal{A}_n$. We consider the node reached at time n as the new root of our tree and take expectation from there. The conditional expectation $E_P(\cdot|\mathcal{A}_n)$ depends on the value of the filtration, it is an $\mathcal{A}_n$-measurable random variable. It is

---

[2]Each of these events, G_i, corresponds to a particular path from S_0 to a node at time n.

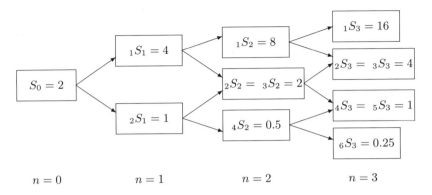

$$n = 0 \qquad n = 1 \qquad n = 2 \qquad n = 3$$

Fig. 9.1 The tree diagram for Example 9.2.

random in the sense that its value depends on the portion of the path from 0 to n which is unknown to us until time n.

Example 9.2. Consider the tree given in Fig. 9.1. Let the probabilities of 'up' and 'down' be equal to $1/2$. Find $E(S_3|\mathcal{A}_2)$.

Answer. We have four different paths by which we can arrive at time $n = 2$. We get

$$\text{First path: } E(S_3|\mathcal{A}_2)(UU) = \frac{1}{2} \cdot 16 + \frac{1}{2} \cdot 4 = 10,$$

$$\text{Second path: } E(S_3|\mathcal{A}_2)(UD) = \frac{1}{2} \cdot 4 + \frac{1}{2} \cdot 1 = 2.5,$$

$$\text{Third path: } E(S_3|\mathcal{A}_2)(DU) = \frac{1}{2} \cdot 4 + \frac{1}{2} \cdot 1 = 2.5,$$

$$\text{Fourth path: } E(S_3|\mathcal{A}_2)(DD) = \frac{1}{2} \cdot 1 + \frac{1}{2} \cdot 0.25 = 0.625.$$

So, $E(S_3|\mathcal{A}_2)$ is a random variable taking values in $\{0.625, 2.5, 10\}$. Further, we can compute probabilities with which it takes these values. Because the probability of each of these four paths is $1/4$, the random variable takes: 0.625 with probability $1/4$, 2.5 with probability $1/2$ and 10 with probability $1/4$. This random variable is $\mathcal{A}_2$-measurable (if we had information up to $n = 2$, we would know which value the random variable takes).

Example 9.3. (*Continuation of Example 9.2*). Now compute the expectation based on the information at time one: $E(E(S_3|\mathcal{A}_2)|\mathcal{A}_1)$.

Answer. There are two paths by which we can arrive at $n = 1$. For the first path we get:

$$E(E(S_3|\mathcal{A}_2)|\mathcal{A}_1)(U) = \frac{1}{2} \cdot E(S_3|\mathcal{A}_2)(UU) + \frac{1}{2} \cdot E(S_3|\mathcal{A}_2)(UD)$$

$$= \frac{1}{2} \cdot 10 + \frac{1}{2} \cdot 2.5 = 6.25;$$

and for the second path:

$$E(E(S_3|\mathcal{A}_2)|\mathcal{A}_1)(D) = \frac{1}{2} \cdot E(S_3|\mathcal{A}_2)(DU) + \frac{1}{2} \cdot E(S_3|\mathcal{A}_2)(DD)$$

$$= \frac{1}{2} \cdot 2.5 + \frac{1}{2} \cdot 0.625 = \frac{25}{16}.$$

The conditional expectation $E(E(S_3|\mathcal{A}_2)|\mathcal{A}_1)$ is a $\mathcal{A}_1$-measurable random variable with values in $\{6.25, 25/16\}$, each taken with probability $1/2$.

Example 9.4. (*Continuation of Examples 9.2 and 9.3*). Compute $E_p(S_3|\mathcal{A}_1)$.

Answer. We can arrive at $n = 1$ via two paths and we obtain:

First path: $E(S_3|\mathcal{A}_1)(U) = \frac{1}{4} \cdot 16 + \frac{1}{2} \cdot 4 + \frac{1}{4} \cdot 1 = 6.25,$

Second path: $E(S_3|\mathcal{A}_1)(D) = \frac{1}{4} \cdot 4 + \frac{1}{2} \cdot 1 + \frac{1}{4} \cdot \frac{1}{4} = \frac{25}{16}.$

We compare Examples 9.3 and 9.4 and observe that $E(E(S_3|\mathcal{A}_2)|\mathcal{A}_1) = E(S_3|\mathcal{A}_1)$ which illustrates the tower property formulated below.

Let us list (without proofs) the important properties of conditional expectations.

The tower property of conditional expectations. *For any $0 < l < k \le n$:*

$$E(E(\xi|\mathcal{A}_{n-l})|\mathcal{A}_{n-k}) = E(\xi|\mathcal{A}_{n-k}). \tag{9.6}$$

This property, which is also known as iterated conditioning, can be interpreted and remembered in the following way. Whilst taking conditional expectations, a smaller filtration 'eats' a greater filtration. Algebra $\mathcal{A}_{n-k}$

is smaller than A_{n-l}, $l < k \leq n$, in the usual sense: at time $n - k$ less information is available than at time $n - l$.

In particular, it follows from the tower property that

$$E(E(\xi|A_n)) = E(\xi). \tag{9.7}$$

Exercise 9.1. Explain why (9.7) follows from (9.6).

Taking out what is known in conditional expectation. *If η is $\mathcal{G}$-measurable, then*

$$E(\eta\xi|\mathcal{G}) = \eta \cdot E(\xi|\mathcal{G}),$$

and, in particular,

$$E(\eta|\mathcal{G}) = \eta.$$

This property can be interpreted as follows. If η is known by time n and we condition it on the information A_n up to time n, we can treat η as a constant: $E(\eta\xi|A_n) = \eta \cdot E(\xi|A_n)$.

Example 9.5. We have

$$E(S_2|A_2) = S_2.$$

Linearity of conditional expectations. *For any constants a and b, we have*

$$E(a\xi + b\eta|\mathcal{G}) = aE(\xi|\mathcal{G}) + bE(\eta|\mathcal{G}).$$

The properties analogous to properties 7 and 8 of conditional expectations given a random variable from Section 8.4 also hold:

- If $\xi \leq \zeta$ then $E(\xi|\mathcal{G}) \leq E(\zeta|\mathcal{G})$.
- $|E(\xi|\mathcal{G})| \leq E(|\xi| \ |\mathcal{G})$.

Now we are going to consider an important class of random processes.

9.2 Martingales

Definition 9.8. *(Discrete case).* A process ζ_n is a **martingale** *with respect to a measure Q (a Q-**martingale**) and a filtration $\{A_i\}_{i \geq 0}$ if*

$$E_Q|\zeta_n| < \infty, \quad E_Q(\zeta_{n+1}|A_n) = \zeta_n \quad \text{for all } n.$$

*The corresponding measure Q is called a **martingale measure**.*

This means that for ζ_n to be a martingale with respect to Q, the future expected value at time $n + 1$ of the process ζ_n under measure Q conditional on the history up until time n is the process' value at time n. Or simply, a martingale is a stochastic process such that the expected value of the process at some future time, given its past history up to now, is equal to the current value. In other words, this means that the process ζ_n has no drift under Q, i.e. no bias up or down in its value under the expectation operator E_Q.

Example 9.6. (*Trivial*). The process which always equals a fixed value is a martingale with respect to all possible measures.

Property 9.1. *If ζ_n is a martingale with respect to Q, then for any $j > 0$*

$$E_Q(\zeta_{m+j}|\mathcal{A}_m) = \zeta_m. \tag{9.8}$$

Proof. Using the tower property of conditional expectations and the fact that ζ_m is a martingale, we obtain

$$
\begin{aligned}
E_Q(\zeta_{m+j}|\mathcal{A}_m) &= E_Q(E_Q(\zeta_{m+j}|\mathcal{A}_{m+j-1})|\mathcal{A}_m) \\
&= E_Q(\zeta_{m+j-1}|\mathcal{A}_m) \\
&= E_Q(E_Q(\zeta_{m+j-1}|\mathcal{A}_{m+j-2})|\mathcal{A}_m) \\
&= E_Q(\zeta_{m+j-2}|\mathcal{A}_m) \\
&= \cdots = \\
&= E_Q(\zeta_{m+1}|\mathcal{A}_m) \\
&= \zeta_m.
\end{aligned}
$$

Example 9.7. We know from Example 8.10 that

$$E_Q(e^{-r\Delta t}S_1) = S_0. \tag{9.9}$$

Looking at (9.9), introduce the *discounted price process*

$$\eta_n := e^{-rn\Delta t}S_n, \tag{9.10}$$

where S_n is defined by our binary tree model (I) of price evolution.

According to (9.9),

$$E_Q\eta_1 = \eta_0.$$

It may be that η_n is a martingale with respect to the measure Q defined as:

$$
\begin{aligned}
Q(\xi_1 = b_1) = q_1 &= \frac{e^{r\Delta t} - (1 + a_1)}{b_1 - a_1}, \\
Q(\xi_1 = a_1) &= 1 - q_1,
\end{aligned}
$$

$$Q[\xi_{n+1} = b_{n+1}|\xi_1, \ldots, \xi_n] = \frac{e^{r\Delta t} - (1 + a_{n+1})}{b_{n+1} - a_{n+1}}, \tag{II}$$

$$Q[\xi_{n+1} = a_{n+1}|\xi_1, \ldots, \xi_n] = \frac{1 + b_{n+1} - e^{r\Delta t}}{b_{n+1} - a_{n+1}}, \quad n = 0, \ldots, N-1.$$

Using properties of conditional expectations, we obtain

$$\begin{aligned}
E_Q(\eta_{n+1}|\mathcal{A}_n) &= E_Q(e^{-r(n+1)\Delta t}S_{n+1}|\mathcal{A}_n) \\
&= e^{-r(n+1)\Delta t}E_Q(S_n + S_n\xi_{n+1}|\mathcal{A}_n) \\
&= e^{-r(n+1)\Delta t}(S_n + S_n E_Q(\xi_{n+1}|\mathcal{A}_n)) \\
&= e^{-r(n+1)\Delta t}S_n(1 + e^{r\Delta t} - 1) \\
&= e^{-rn\Delta t}S_n = \eta_n.
\end{aligned}$$

The condition $E_Q|\eta_n| < \infty$ is evidently satisfied.

Conclusion: the process $\eta_n = e^{-rn\Delta t}S_n$ is a martingale with respect to the measure Q from (II).

We will see very soon that this Q is (as you might have already guessed) the risk-neutral measure for the multiperiod binary model and, moreover, that a risk-neutral measure can be defined as a measure under which the discounted stock price is a martingale. The key fact is that under risk-neutral measure, for stock that pays no dividend, the best estimate based on the information at time n of the value of the discounted stock price at time $n+1$ is the discounted stock price at time n. We can also conclude that under the risk-neutral measure the mean rate of return for the stock is r, the same as the rate of return for the money market.

Remark 9.2. Q-dependence in Definition 9.8 is essential because exactly the same process can be a martingale with respect to one measure and not to another.

Exercise 9.2. Illustrate the above remark by an example considering a stochastic process which is a martingale under one measure but not under another.

The theory of martingales is one of the main mathematical tools in Financial Mathematics. David Williams (2007) FRS (Fellow of the Royal Society), notes:

'. . . at one stage, even being able to say the word "martingale" tends to get you a highly paid job in finance. Nowadays, everyone in finance knows something about martingales, so you would need to, too.'

Example 9.8. (*Creating martingales by 'projection'*). Suppose we have a filtration $\{\mathcal{A}_i\}$ and a random variable ξ on $(\Omega, \mathcal{A}, P)$ with $E|\xi| < \infty$. The process

$$\zeta_n := E(\xi|\mathcal{A}_n), \tag{9.11}$$

is a martingale. Indeed, by the tower property of conditional expectation, we get

$$\zeta_n \overset{(9.11)}{=} E(\xi|\mathcal{A}_n) \overset{\text{tower property}}{=} E(E(\xi|\mathcal{A}_{n+1})|\mathcal{A}_n) \overset{(9.11)}{=} E(\zeta_{n+1}|\mathcal{A}_n).$$

It is also not difficult to see that $E|\zeta_n| < \infty$:

$$E|\zeta_n| = E|E(\xi|\mathcal{A}_n)| \le EE(|\xi|\,|\mathcal{A}_n)| = E|\xi| < \infty.$$

Example 9.9. (*Binomial representation theorem*[3]). Consider the binary tree model (I). Assume that the measure Q is such that the discounted price process η_n from (9.10) is a Q-martingale. Let M_n be any other Q-martingale. Show that there exists an $\mathcal{A}_n$-adapted process θ_n such that

$$M_n = M_0 + \sum_{i=1}^{n} \theta_{i-1}\,(\eta_i - \eta_{i-1}), \tag{9.12}$$

i.e. we consider two martingales, η_i and M_i, and would like to prove that they are related via (9.12).

Answer. The increments over the branch of the processes η_n and M_n are the random variables:

$$\Delta\eta_i = \eta_i - \eta_{i-1}, \quad \Delta M_i = M_i - M_{i-1}.$$

Due to the tree structure, the filtration $\mathcal{A}_i$ has two choices beyond $\mathcal{A}_{i-1}$ corresponding to the 'up' movement and 'down' movement. Consider a single branching from a node at time $i - 1$ to two nodes 'up' and 'down' at time i. For a fixed $\mathcal{A}_{i-1}$, there are only two places to go: η_i can take only two values, $_U\eta_i$ and $_D\eta_i$, and M_i can take only two values, $_U M_i$ and $_D M_i$. After we fix $\mathcal{A}_{i-1}$, we know the values of η_{i-1}, $_U\eta_i$, $_D\eta_i$, M_{i-1}, $_U M_i$ and $_D M_i$. Let the width for η_i be $\delta\eta_i = \;_U\eta_i - \;_D\eta_i$ and for M_i be $\delta M_i = \;_U M_i - \;_D M_i$, both of them dependent on $\mathcal{A}_{i-1}$ only – they are measurable with respect to $\mathcal{A}_{i-1}$. Then the ratio of these branch widths is

$$\theta_{i-1} = \frac{\delta M_i}{\delta\eta_i}.$$

---

[3] A continuous-time analogue of the binomial representation theorem is the martingale representation theorem (see, e.g., Etheridge, 2002; Baxter and Rennie, 1996; Karatzas and Shreve, 1998; Shreve, 2004); however, we do not consider it in this course.

Let the offset for η_i be $\gamma_{\eta_i} = \left(_U\eta_i + {_D}\eta_i\right)/2 - \eta_{i-1}$ hence $\eta_i = \eta_{i-1} + \gamma_{\eta_i} \pm \delta\eta_i/2$. Analogously, we can introduce the offset γ_{M_i} for M_i. Within our set-up, the offsets depend on $\mathcal{A}_{i-1}$ only. Any random variable dependent on the branch (i.e. on $\mathcal{A}_{i-1}$) is fully determined by its width size δ and a constant offset γ. Thus, if we would like to construct one process out of another, we, in general, base such a construction on a scaling (to match the width) and a shift (to match the offsets). Taking into account the relative shift $\lambda = \gamma_{M_i} - \theta_{i-1}\gamma_{\eta_i}$, we can write

$$\Delta M_i = \theta_{i-1}\Delta\eta_i + \lambda.$$

Here, the relative shift λ is also dependent on $\mathcal{A}_{i-1}$ only. However, we assumed that η_n and M_n are martingales, then $E_Q(\Delta\eta_i|\mathcal{A}_{i-1}) = 0$ and $E_Q(\Delta M_i|\mathcal{A}_{i-1}) = 0$. Because θ_{i-1} and λ are measurable with respect to $\mathcal{A}_{i-1}$, we get

$$E_Q(\Delta M_i|\mathcal{A}_{i-1}) = \theta_{i-1}E_Q(\Delta\eta_i|\mathcal{A}_{i-1}) + \lambda,$$

which implies that $\lambda = 0$. We conclude that we have found such $\mathcal{A}_{i-1}$-measurable θ_{i-1} that

$$M_i - M_{i-1} = \theta_{i-1}\left(\eta_i - \eta_{i-1}\right). \tag{9.13}$$

Finally, summing (9.13) from $i = 1$ to n and observing that on the left we obtain a telescopic sum, we arrive at (9.12).

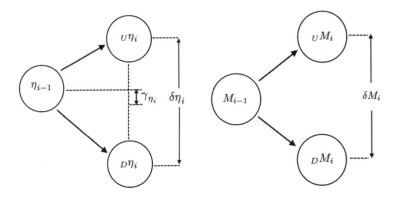

Fig. 9.2   Illustration for Example 9.9.

Historical comment. The term 'martingale' came to Probability theory from gambling. As mentioned in Davis and Etheridge (2006), Casanova, in his memoirs, wrote about winning a fortune at the roulette table playing a martingale, only to lose it a few days later. The 'infallible martingale'[4] is a gambling strategy aimed at achieving a sure profit in a game like roulette by making a sequence of independent bets such that one puts £x on their number of the day in roulette and keeps doubling the stake until the number wins[5]. When this happens, the win not only covers all the previous losses but also gives a profit. For this strategy the odds are not important assuming that a win eventually happens. 'But one must play for a very long time, for many days, being satisfied with little when no luck, and not violently rushing upon a chance' wrote F. M. Dostoevsky[6], who was addicted to gambling for many years, to his wife on 6 May 1867. The 'infallible martingale' strategy is, in fact, not infallible because time is finite as well as the initial capital of any player. So, do not try it yourself! The word 'martingale' has other meanings as well: a strap attached to a fencer's épée, a strut under the bowsprit of a sailing boat or the strap of a horse's harness connecting the girth to the noseband and preventing the horse from throwing back its head. The latter, like in the gambling strategy, allows free movement in one direction but not in the other. Jean Ville (French mathematician, 1910–1988) introduced the notion of a martingale in Probability theory in 1939, he used it to describe the fortune of a player in a fair game rather than the gambling strategy. Much of the initial development of the theory of martingales was done by Joseph Leo Doob (American mathematician, 1910–2004) in the 1940s. It should also be noted that roulette is not a fair game (it is biased in favour of the casino) and the player's fortune is actually a *supermartingale* (see definition in Section 12.2).

9.3   Change of Measure and (Discrete) Radon–Nikodym Derivative

As we observed before, there are two types of probability measures that can be considered within our price evolution model. One of them is usually called the actual, real or market probability measure, by which one means the measure which corresponds to empirical estimation of the model parameters (like we do in statistics and econometrics, which we will briefly

---

[4]There are also other 'martingale' strategies in gambling.

[5]In application to the game of tossing a fair coin this strategy is related to the famous St. Petersburg paradox.

[6]A free translation from the Russian collection of F. M. Dostoevsky's works, Volume 15, Letters (Nauka, Moscow, 1996), p. 302.

discuss in Section 14.2). The other is the risk-neutral measure, under which the discounted prices are martingales (see Example 9.7 and the more precise definition in Section 9.4). These two probability measures give different weights to the price paths. They agree on which price paths are possible (i.e. which paths have positive probability of occurring); however, they disagree only on what these positive probabilities are. Such measures are called *equivalent*[7]. We can regard the measure Q as a re-weighting of the measure P.

Consider a finite sample space Ω on which we have two probability measures P and Q. For simplicity, we assume that both P and Q give positive probability to every element $\omega \in \Omega$, so that we can form the quotient

$$Z(\omega) = \frac{Q(\omega)}{P(\omega)}.$$

This Z is obviously a random variable. It is called the *Radon–Nikodym derivative* of Q with respect to P, although in the discrete case it is really a quotient rather than a derivative.

This random variable has three important properties:

1. $P(Z > 0) = 1$.
2. $E_P Z = 1$.
3. For any random variable ξ

$$E_Q \xi = E_P (Z\xi). \tag{9.14}$$

In our case, the first property follows from the fact that we assumed that $Q(\omega) > 0$ for all ω. The third property is proved as follows:

$$E_Q \xi = \sum_{\omega \in \Omega} \xi(\omega) Q(\omega) = \sum_{\omega \in \Omega} \xi(\omega) \frac{Q(\omega)}{P(\omega)} P(\omega)$$
$$= \sum_{\omega \in \Omega} \xi(\omega) Z(\omega) P(\omega) = E_P (Z\xi).$$

The second property immediately follows from the third with $\xi = 1$.

Let us now consider the measurable space $(\Omega, \mathcal{A})$ and the filtration $\mathcal{A}_n$ associated with our price model (I). Introduce the *Radon–Nikodym derivative process* as

$$Z_n = E_P(Z|\mathcal{A}_n). \tag{9.15}$$

---

[7]Consequently, in Section 9.4 we use the term 'an equivalent martingale measure (EMM)' on an equal footing with 'risk-neutral measure'.

Due to Example 9.8, Z_n is a martingale under the measure P. We note that $Z_0 = 1$ (see property 2 of Radon–Nikodym derivatives) and $Z_N = Z$ (recall that we consider N time steps in (I) and, consequently, $\mathcal{A} = \mathcal{A}_N$).

Let ζ_n be an $\mathcal{A}_n$-adapted process. According to property 3 of Radon–Nikodym derivatives and properties of conditional expectations, we obtain

$$E_Q(\zeta_n) \overset{\text{property 3}}{=} E_P(Z\zeta_n) \overset{\text{Tower property}}{=} E_P(E_P(Z\zeta_n|\mathcal{A}_n)) \qquad (9.16)$$

$$\overset{\zeta_n \in \mathcal{A}_n}{=} E_P(\zeta_n E_P(Z|\mathcal{A}_n)) \overset{(9.15)}{=} E_P(Z_n\zeta_n).$$

One might notice that the process of changing the martingale measure can be viewed as a re-weighting of the probabilities of paths under the original measure P according to a positive P-martingale with mean being equal to one[8].

In addition, one can prove the formula relating conditional expectations:

$$E_Q(\zeta_n|\mathcal{A}_m) = \frac{1}{Z_m} E_P(Z_n\zeta_n|\mathcal{A}_m). \qquad (9.17)$$

9.4 Application of Martingales and First Fundamental Theorem of Asset Pricing

In this section we do not restrict the price process S_n to follow the model (I). We consider a filtered probability space $(\Omega, \mathcal{A}, \{\mathcal{A}_i\}, P)$ and the price process S_n defined on it. Here P is an 'actual' probability measure[9]. Recall that the filtration $\mathcal{A}_n$ is interpreted in our case as the information about the price behaviour from time 0 to n. As before, we also suppose that there is a bank account B_n with a known fixed continuously compounded interest rate r. Let P and Q be equivalent measures[10] (see Section 9.3) and let the discounted price process

$$\tilde{S}_n := e^{-rn\Delta t} S_n \qquad (9.18)$$

be a Q-martingale. Such a measure Q is called **an equivalent martingale measure (EMM)** or **a risk-neutral measure**. For instance, we know

---

[8]This procedure of re-weighting according to a positive martingale can be extended to the continuous case. Change of measure in the continuous case (in particular, Girsanov's theorem) plays the crucial role in pricing derivatives (see, e.g., Baxter and Rennie, 1996; Lamberton and Lapeyre, 2007; Shreve, 2004; Etheridge, 2002).

[9]In particular, via $P(\omega) > 0$ it tells us which paths are 'practically' possible in the price model.

[10]In the case of the model (I) it is equivalent to the requirement that $Q(\omega) > 0$ for all $\omega \in \Omega$ (i.e. that each path in the model (I) has a positive probability), where Ω is the sample space corresponding to the 'experiment' (I), in which we already assumed that $P(\omega) > 0$ for all $\omega \in \Omega$ (see the beginning of Chapter 7).

from Example 9.7 that if S_n is from (I) then Q defined in (II) satisfies this requirement.

We are about to discover the important consequences of the existence of EMM for financial applications. To this end, we introduce the portfolio sequence (ϕ_n, ψ_n), $n = 0, \ldots, N - 1$, where ϕ_n is the amount of stock held by the writer during the $(n + 1)$ period and ψ_n is the amount of money at the bank account deposited by the writer at time n. It is natural from the applicable point of view to consider only such strategies (ϕ_n, ψ_n) that the process (ϕ_n, ψ_n) is $\mathcal{A}_n$-adaptive. In other words, the portfolio (ϕ_n, ψ_n) is formed at time n using the information about the market behaviour up to time n only, it does not use information about the 'future'. Such strategies are called *admissible*. We also assume that the random variables ϕ_n and ψ_n are bounded, which is a natural assumption in the considered discrete set-up and also natural from the applicable point of view.

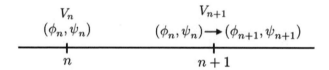

Fig. 9.3   Admissible self-financing strategy.

Suppose a portfolio (ϕ_n, ψ_n) was formed at time n when it costed

$$V_n = \phi_n S_n + \psi_n. \tag{9.19}$$

We arrive with this portfolio at time $n+1$ and its value at time $n+1$ becomes

$$V_{n+1} = \phi_n S_{n+1} + \psi_n e^{r\Delta t}. \tag{9.20}$$

Further, at the time $n + 1$, we can use the wealth V_{n+1} to form a new portfolio (ϕ_{n+1}, ψ_{n+1}) (remember that we use self-financing strategies (see Definition 7.1)) and so

$$V_{n+1} = \phi_{n+1} S_{n+1} + \psi_{n+1}. \tag{9.21}$$

Example 9.10. Show that the discounted value of the portfolio (i.e. the discounted wealth)

$$\tilde{V}_n := e^{-rn\Delta t} V_n \tag{9.22}$$

is also a martingale under an EMM Q.

Solution. Because (cf. (9.20) and (9.22)):

$$\tilde{V}_{n+1} = \phi_n \tilde{S}_{n+1} + e^{-rn\Delta t} \psi_n,$$

we have

$$E_Q(\tilde{V}_{n+1}|\mathcal{A}_n) = E_Q(\phi_n \tilde{S}_{n+1} + e^{-rn\Delta t} \psi_n|\mathcal{A}_n).$$

Using properties of conditional expectations and the facts that (ϕ_n, ψ_n) is $\mathcal{A}_n$-measurable and $\tilde{S}_n$ is a martingale under Q, we obtain

$$E_Q(\phi_n \tilde{S}_{n+1} + e^{-rn\Delta t} \psi_n|\mathcal{A}_n) = \phi_n E_Q(\tilde{S}_{n+1}|\mathcal{A}_n) + e^{-rn\Delta t}\psi_n$$

$$\overset{(9.18)}{=} \phi_n \tilde{S}_n + e^{-rn\Delta t}\psi_n$$

$$\overset{(9.21)\text{ at time }n\text{ and }(9.22)}{=} \tilde{V}_n.$$

Hence,

$$\tilde{V}_n = E_Q(\tilde{V}_{n+1}|\mathcal{A}_n). \tag{9.23}$$

The condition $E_Q|\tilde{V}_k| < \infty$, $k = 0, \ldots, N$, is satisfied because $\tilde{S}_n$ is a Q-martingale and ϕ_n and ψ_n are bounded. We summarise this result as the following proposition.

Proposition 9.1. *If there is an EMM Q then for any $\mathcal{A}_n$-adaptive strategy (ϕ_n, ψ_n) the discounted wealth process $\tilde{V}_n$ is a Q-martingale.*

Again pay attention to the fact that we did not use a particular model for S_n here; we only required that $\tilde{S}_n$ is a Q-martingale.

Due to the martingale property (9.8) from Section 9.2, we have

$$E_Q \tilde{V}_N = V_0. \tag{9.24}$$

The consequence of (9.24) is that there is no arbitrage in the market model under consideration. Indeed, if there were an arbitrage, an investor could start with $V_0 = 0$ and find a portfolio process (i.e. a strategy) whose corresponding wealth process $V_1, V_2, \ldots, V_N$ satisfies $V_N \geq 0$ for all paths $\omega \in \Omega$ and $V_N(\omega') > 0$ for at least one path $\omega' \in \Omega$ with $P(\omega') > 0$ and hence with $Q(\omega') > 0$. However, then we would have $V_0 = 0$ and $E_Q \tilde{V}_N > 0$ which is impossible since it contradicts (9.24). Thus, we have proved the following proposition.

Proposition 9.2. *If there is an EMM then there is no arbitrage in the market model.*

This proposition can be generalised for a market with many underliers. In general, if there is a risk-neutral measure in a price model (i.e. a measure that agrees with the 'actual' probability measure about which price paths have zero probability and under which the discounted prices of all primary assets are martingales), then there is no arbitrage in the model. A proof of such a theorem is essentially as above: under a risk-neutral measure, the discounted wealth process has a constant expectation, so it cannot begin at zero and later be strictly positive with positive probability unless it also has a positive probability of being strictly negative.

The opposite statement to Proposition 9.2, i.e. 'absence of arbitrage implies the existence of an EMM', is also valid. Its proof can be found, e.g., in Pliska (1997) but we skip it here. The theorem about equivalence of the existence of a risk-neutral measure and absence of arbitrage is often called the *first fundamental theorem of asset pricing*[11].

First fundamental theorem of asset pricing. *There is no arbitrage in the market model if and only if there is an EMM.*

9.5 Uniqueness of Arbitrage Price and Replicating Strategy

As in the previous section, we assume existence of an EMM and do not restrict S_n to the model (I) but we also require that at each time step n at least for a single $\omega \in \Omega$ with $Q(\omega) > 0$ the discounted price process is such that

$$\tilde{S}_{n+1}(\omega) \neq \tilde{S}_n(\omega). \tag{9.25}$$

This non-degeneracy condition is natural for risky assets and, for example, consistent with the condition (Ib) in our model (I).

Suppose that a writer sells a European-type option with maturity time N and a payoff f which is $\mathcal{A}_N$-measurable and that he wants the value of his portfolio at time N to meet the obligation:

$$V_N = f$$

for any realisation of the price process S_n. We assume that $E_Q|f| < \infty$.

According to (9.24), if there is an admissible strategy allowing the writer to meet this obligation f (i.e. there is a sequence (ϕ_n, ψ_n), $n = 0, \ldots, N-1$,

---

[11]Although there are additional technicalities involved, this fundamental theorem has essentially the same statement for continuous-time models.

such that $V_N = f$ on all realisations of trajectories S_n, $n = 0, \ldots, N$) then his initial wealth should be

$$x = V_0 = E_Q \tilde{V}_N = e^{-rT} E_Q f, \qquad (9.26)$$

which gives us an arbitrage price of the option. Such a strategy is called a *replicating strategy* because it replicates the contingent claim.

Let us show that this price is unique. We will prove it via contradiction. Suppose there are two different prices, $x^{(1)}$ and $x^{(2)}$, and two corresponding strategies $(\phi_n^{(i)}, \psi_n^{(i)})$, $i = 1, 2$, such that the corresponding wealths at time N meet the obligation: $V_N^{(i)} = f$, $i = 1, 2$, for any realisation of the price process S_n. Recalling that the discounted wealths $\tilde{V}_n^{(i)}$ are Q-martingales (see Proposition 9.1) and the property of martingales (9.8), we get

$$\tilde{V}_n^{(i)} = E_Q(\tilde{V}_N^{(i)} | \mathcal{A}_n). \qquad (9.27)$$

However, we also require $V_N^{(1)} = V_N^{(2)}$, which together with (9.27) implies that $\tilde{V}_n^{(1)} = \tilde{V}_n^{(2)}$ for all $n = 0, \ldots, N$, and, in particular, that $x^{(1)} = V_0^{(1)}$ and $x^{(2)} = V_0^{(2)}$ coincide.

Further, it also follows from this consideration that the corresponding strategy is unique. Indeed, we obtain from (9.20)–(9.21):

$$\tilde{V}_{n+1}^{(i)} - \tilde{V}_n^{(i)} = \phi_n^{(i)} \left(\tilde{S}_{n+1} - \tilde{S}_n \right), \quad i = 1, 2.$$

But we have already shown that $\tilde{V}_n^{(1)} = \tilde{V}_n^{(2)}$ and, consequently, $\tilde{V}_{n+1}^{(1)} - \tilde{V}_n^{(1)} = \tilde{V}_{n+1}^{(2)} - \tilde{V}_n^{(2)}$. Due to the non-degeneracy condition (9.25), the random variable $\tilde{S}_{n+1} - \tilde{S}_n$ at each step n cannot be identically equal to zero. Therefore, $\phi_n^{(1)} = \phi_n^{(2)}$ for all n. Then, obviously (*Exercise.* Please check this.), $\psi_n^{(i)}$, $i = 1, 2$, also coincide. Thus, $(\phi_n^{(1)}, \psi_n^{(1)}) = (\phi_n^{(2)}, \psi_n^{(2)})$ for all n. We have arrived at the following result.

Proposition 9.3. *If there is an EMM Q and there is an $\mathcal{A}_n$-adaptive replicating strategy (ϕ_n, ψ_n) for a European option with payoff f being $\mathcal{A}_N$-measurable then the arbitrage price of this option*

$$x = e^{-rT} E_Q f \qquad (9.28)$$

is unique and under the non-degeneracy condition (9.25) the replicating strategy (ϕ_n, ψ_n) is also unique.

Remark 9.3. Note that in the pricing formula (9.28) Q can be any EMM, i.e. the statement of Proposition 9.3 and, in particular, the arbitrage price x from (9.28) does not depend on the choice of the EMM, even if there is more than one EMM (see also Chapter 11).

In connection with Proposition 9.3, let us introduce the term 'attainable'.

Definition 9.9. *A contingent claim f due at the maturity N is **attainable** if there exists an admissible strategy which replicates, or hedges, f, i.e. the wealth process V_n corresponding to this strategy satisfies $V_N = f$.*

The results obtained in this section are valid for any claim f which is $\mathcal{A}_N$-measurable. In particular, it is valid for claims of the form $f(S_N)$, i.e. which depends on the price of stock at the final moment only (like the plain vanilla European calls and puts we have already considered). However, the results are also true for *path-dependent options* of European-type (e.g., for European-type *Asian options*). To this end, consider the following example.

Example 9.11. A European-type arithmetic-average *Asian call option* has the payoff which depends on the path:

$$f_N = \left(\frac{1}{N+1} \sum_{n=0}^{N} S_n - K \right)_+$$

and which can be exercised at the maturity time N only. A European-type arithmetic-average *Asian put* has the payoff

$$f_N = \left(K - \frac{1}{N+1} \sum_{n=0}^{N} S_n \right)_+ .$$

These are examples of *exotic options* (see also e.g., Avellaneda and Laurence, 2000; Bingham and Kiesel, 2004; Kolb, 2003; Hull, 2003). An investor might be interested in Asian options due to the fact that their performance is less affected by a possible market manipulation of the underlying at the maturity than the performance of European options depending on the state of the market at the maturity only.

Chapter 10

Multiperiod Binary Tree Model

'On a binomial tree, prices move like knights on a chessboard, one discrete step forward in time and up or down a notch in price. Binomial trees are easy to draw and, in a jerky way, mimic the behavior of real prices or indices.' (Derman, 2004, pp. 234–235)

In Sections 9.4 and 9.5 we did not need to specify a particular model for the price process S_n and the results were of a rather general nature. Now we will be more specific and return to considering European-type claims of the form $f(S_N)$ with the underlier following our binary tree model (I). We note that similar arguments to the ones considered in this chapter can be applied to more general European-type derivatives with the underlier modelled by (I) like path-dependent options from Example 9.11.

Let us consider the case of an *arbitrary* N in the binary model (I) illustrated in Fig. 10.1.

We recall that in the case of the model (I) the measure Q defined in (II) is an EMM and there is no arbitrage in the corresponding market (see the condition (Ic) and also the first fundamental theorem of asset pricing).

10.1 Backward Induction and the Existence of Hedging Strategy

Now our aim is to construct an $\mathcal{A}_n$-adaptive replicating strategy (ϕ_n, ψ_n) for a European option with payoff $f(s)$ within the framework of the model (I). If we succeed then Proposition 9.3 will imply that the arbitrage price from (9.28) is unique as well as the replicating strategy (remember that we already showed in Example 9.7 that the measure Q from (II) is an EMM).

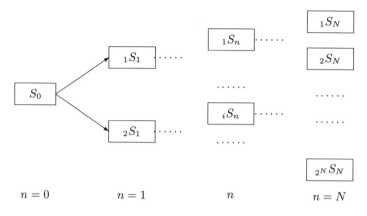

Fig. 10.1   Multiperiod binary tree.

To get a hedging strategy (which in our context means a self-financing strategy which replicates the claim), we go backwards in time. At each step of our consideration we use the one-step (single-period) arguments from Section 7.2.

At time N, the value of the portfolio (ϕ_{N-1}, ψ_{N-1}) formed at time $N-1$ must meet the obligation:

$$V_N = S_N \phi_{N-1} + e^{r\Delta t}\psi_{N-1} = f(S_N).$$

This obligation is translated backwards in time, step by step. Indeed, let us fix some n among $0 \leq n \leq N-1$ and the history $\mathcal{A}_n$ up to time n. Then the future price S_{n+1} and obligation V_{n+1} can take only two values: $_US_{n+1}$ and $_DS_{n+1}$, and $_UV_{n+1}$ and $_DV_{n+1}$, respectively[1]. At this time n, we have to form the portfolio (ϕ_n, ψ_n) so that the obligation V_{n+1} is always met at the next step:

$$\phi_n\, _US_{n+1} + e^{r\Delta t}\psi_n = \, _UV_{n+1} \qquad (10.1)$$
$$\phi_n\, _DS_{n+1} + e^{r\Delta t}\psi_n = \, _DV_{n+1}.$$

Note that at the first step of the backward construction $n = N-1$ and $V_{n+1} = V_N = f(S_N)$. Recall that due to our stock price model (I):

$$_US_{n+1} = S_n(1 + b_{n+1}) \qquad (10.2)$$
$$_DS_{n+1} = S_n(1 + a_{n+1}).$$

---

[1]To be more precise, we should include the dependence on $\mathcal{A}_n$ in the notation but we omit it in order not to overcomplicate the writing; this simplification of the writing should not lead to any confusion.

The equations (10.1) and (10.2) imply (cf. Theorem 7.1):

$$\phi_n = \frac{{}_U V_{n+1} - {}_D V_{n+1}}{S_n(b_{n+1} - a_{n+1})},$$ (10.3)

$$\psi_n = e^{-r\Delta t}[{}_U V_{n+1} - \frac{{}_U V_{n+1} - {}_D V_{n+1}}{b_{n+1} - a_{n+1}}(1 + b_{n+1})],$$

and the value of the portfolio at the time n should be:

$$V_n = \phi_n S_n + \psi_n,$$ (10.4)

which translates the obligation to the next step. We also observe that Example 9.10 implies that (also see Theorem 8.1):

$$V_n = e^{-r\Delta t} E_Q(V_{n+1}|\mathcal{A}_n) = e^{-r\Delta t(N-n)} E_Q(V_N|\mathcal{A}_n),$$

where Q is from (II). In our consideration, the n was an arbitrary moment between 0 and $N - 1$, so if we choose the portfolio according to (10.3) at each step of the binary tree, we replicate the claim.

To conclude, we have constructed the replicating strategy and all the conditions of Proposition 9.3 hold, then the price

$$x = V_0 = e^{-rT} E_Q f(S_N),$$ (10.5)

with the risk-free probability Q defined in (II) is the unique arbitrage price for this option, and the replicating strategy (ϕ_n, ψ_n), $n = N - 1, \ldots, 1, 0$, from (10.3) is unique as well. Further, we also observe that any European-style derivative claim can be replicated on the binary tree model (I) by trading with a self-financing portfolio.

Remark 10.1. We have constructed the replicating strategy following the algorithmic arguments. However, it can also be obtained using the binomial representation theorem from Example 9.9 (see, e.g., pp. 36–39 in Baxter and Rennie, 1996). In the case of continuous time the algebraic arguments we used here are not possible and the key for the construction of hedging strategies is the martingale representation theorem which is a continuous-time analogue of the binomial representation theorem.

Remark 10.2. Note that if we already computed V_n and ϕ_n then (10.4) implies

$$\psi_n = V_n - \phi_n S_n.$$ (10.6)

Remark 10.3. Using the Radon–Nikodym derivative Z introduced in Section 9.3, we can express the option price as expectation with respect to the 'market' probability P:

$$x = e^{-rT} E_P[Zf(S_N)].$$

10.2   Algorithm for the Writer

At this stage, it is appropriate to discuss delta, an important quantity in the pricing and hedging of options. The **delta** of an option is the ratio:

$$\Delta := \frac{\text{change in the price of the option}}{\text{change in the price of the underlying stock}}.$$

In our case,

$$\Delta_{n+1} = \frac{{}_U V_{n+1} - {}_D V_{n+1}}{{}_U S_{n+1} - {}_D S_{n+1}}.$$

The delta Δ_{n+1} is an $\mathcal{A}_n$-measurable random variable. It can be seen (cf. (10.3)) that the delta is the number of stock units ϕ_n the writer should hold in order to create a riskless hedge according to the backward induction. Note that it can be viewed as a finite difference for a change of the portfolio's value with respect to the change of the stock's price. The construction of a riskless hedge is sometimes referred to as **delta hedging**.

     Due to the result obtained via the backward induction in the previous section, the writer should follow the hedging strategy with the initial value x from (10.5), which is presented in its algorithmic form below.

HS: **The hedging strategy (algorithm for the writer)**

1. Set $n = 0$ (start the 'clock') and charge the premium

$$x = V_0 = e^{-rT} E_Q f(S_N),$$

     where Q is the risk-free measure from (II).

2.(a) Evaluate the delta

$$\Delta_{n+1} = \frac{{}_U V_{n+1} - {}_D V_{n+1}}{S_n(b_{n+1} - a_{n+1})}$$

     (it depends on the information $\mathcal{A}_n$);

  (b) Using the existing funds V_n:

     • buy $\phi_n = \Delta_{n+1}$ units of stock (S_n each);
     • put $\psi_n = V_n - \phi_n S_n$ on the bank account.

3. (a) Arrive at next time, $n + 1$, and observe the value of S_{n+1} from the market;

  (b) Revaluate the wealth: $V_{n+1} = \phi_n S_{n+1} + \psi_n e^{r\Delta t}$;

  (c) Redefine $n := n + 1$;

  (d) If $n < N$ go to 2; otherwise, at the maturity time $T = N\Delta t$, the portfolio has the value V_N which matches the payoff function $f(S_N)$.

We conclude that if at $n = 0$ the writer charges for the option the amount of money x from (10.5) then he can replicate the claim. We see that this result is consistent with what we found for $N = 1$ in Section 7.2. We have shown through the above algorithm, HS, that within our model (I) any European-type claim can be replicated with some initial wealth or, in other words, (perfectly) hedged. We summarise the main result of this section in the following theorem.

Theorem 10.1. *Consider the binary model (I). Suppose the conditions (FA) hold. Then there exists a unique arbitrage price x for the European option with payoff function $f(s)$ and maturity time $T = N\Delta t$:*

$$x = e^{-rT} E_Q f(S_N), \tag{10.7}$$

where Q is the EMM defined in (II), and the unique associated hedging strategy defined in the algorithm HS.

We will call (10.7) the **discrete Black–Scholes (BS) formula**.

Remark 10.4. It is easy to generalise Theorem 10.1 to any claim f which is $\mathcal{A}_N$-measurable (see Example 9.11 of European-type path-dependent options). Indeed, due to (9.26), the arbitrage price is found as

$$x = V_0 = E_Q \tilde{V}_N = e^{-rT} E_Q f.$$

A modification of the algorithm HS to European-type path-dependent options is also straightforward.

Let us illustrate the backward induction of Section 10.1 and the hedging strategy HS by considering a particular two-period example, more examples are available in Chapter 13.

Example 10.1. Consider the two-step binary tree model with the structure given in Fig. 10.2.

Let $\Delta t = 1$ and the continuously compounded interest rate r for a single time step Δt satisfy $e^{r\Delta t} = 4/3$.

1. Are there any arbitrage opportunities? If so, find them. If not, go to 2.
2. Consider a European call option with the strike price $K = £30$ and maturity $T = 2\Delta t$.

   (a) Find the premium (the value of the option at $t = 0$).
   (b) Find the hedging portfolio at $t = 0$.

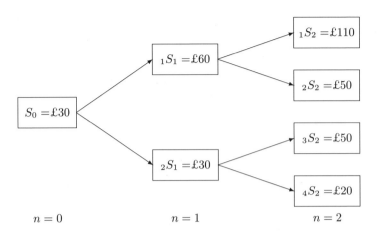

$n = 0$ $n = 1$ $n = 2$

Fig. 10.2   The two-step binary tree model from Example 10.1.

(c) The writer follows the hedging strategy. What should he do at time $t = 1$ if the stock price at $t = 1$ were £30?

Answer. 1. Note that all prices are positive, so the positiveness condition (Ib) is automatically satisfied. Now we have to find the parameters a_n and b_n of the model (I). For the first step, we find a_1 and b_1 from the relations:

$$60 = 30(1 + b_1), \quad 30 = 30(1 + a_1),$$

consequently,

$$a_1 = 0, \quad b_1 = 1.$$

We check the arbitrage-free condition (Ic):

$$1 + a_1 < e^{r\Delta t} < 1 + b_1 \implies 1 < \frac{4}{3} < 2,$$

the inequality is true and, hence, there is no arbitrage present in the first step.

Consider the second step. We have

$$_1 b_2 = \frac{5}{6}, \quad _1 a_2 = -\frac{1}{6},$$

$$_2 b_2 = \frac{2}{3}, \quad _2 a_2 = -\frac{1}{3}.$$

We check that the arbitrage-free condition (Ic) is satisfied at this step as well:

$$1 + {_i}a_2 < e^{r\Delta t} < 1 + {_i}b_2, \quad i = 1, 2;$$

$$\text{for } i = 1 : \quad \frac{5}{6} < \frac{4}{3} < \frac{11}{6};$$

$$\text{for } i = 2 : \quad \frac{2}{3} < \frac{4}{3} < \frac{5}{3}.$$

Thus, we conclude there are no arbitrage opportunities in this market model.

2. To make calculations, we need to find the implied probabilities. Recall that the implied probability to go 'up' in each 'block' is equal to

$$q = \frac{e^{r\Delta t} - (1 + a)}{b - a}.$$

Then, the implied probabilities at the first step are

$$q_1 = \frac{\frac{4}{3} - (1 + 0)}{1} = \frac{1}{3} \text{ ('up')}, \quad 1 - q_1 = \frac{2}{3} \text{ ('down')}$$

and the implied probabilities at the second step are

$$\text{'upper block': } {_1}q_2 = \frac{\frac{4}{3} - (1 - \frac{1}{6})}{\frac{5}{6} + \frac{1}{6}} = \frac{1}{2} \text{ ('up')}, \quad 1 - {_1}q_2 = \frac{1}{2} \text{ ('down')};$$

$$\text{'lower block': } {_2}q_2 = \frac{\frac{4}{3} - (1 - \frac{1}{3})}{\frac{2}{3} + \frac{1}{3}} = \frac{2}{3} \text{ ('up')}, \quad 1 - {_2}q_2 = \frac{1}{3} \text{ ('down')}.$$

The writer's obligation at time $T = 2$ is

$$f(S_2) = (S_2 - K)_+ = (S_2 - 30)_+$$

or

$${_1}f_2 = 80, \quad {_2}f_2 = 20, \quad {_3}f_2 = 20, \quad {_4}f_2 = 0.$$

(2.a) Using the values computed above, we find which value the portfolio should have after the first time step[2]:

$${_1}V_1 = e^{-r\Delta t} E_Q[f(S_2)|S_1 = {_1}S_1] = \frac{3}{4}\left[80 \times \frac{1}{2} + 20 \times \frac{1}{2}\right] = \frac{75}{2} \text{ (£)},$$

$${_2}V_1 = e^{-r\Delta t} E_Q[f(S_2)|S_1 = {_2}S_1] = \frac{3}{4}\left[20 \times \frac{2}{3} + 0 \times \frac{1}{3}\right] = 10 \text{ (£)}.$$

---

[2]Here we use (9.23), which implies that $V_n = e^{-r\Delta t} E_Q(V_{n+1}|\mathcal{A}_n)$, as well as Definitions 9.6 and 8.11.

Finally, we calculate the value of the option at time $t = 0$:

$$x = V_0 = e^{-r\Delta t} E_Q V_1 = \frac{3}{4}\left[\frac{75}{2} \times \frac{1}{3} + 10 \times \frac{2}{3}\right] = \frac{115}{8} \ (\pounds).$$

Thus, the premium is $\pounds\dfrac{115}{8}$.

(2.b) Now we find the hedging portfolio at $t = 0$ (see (10.3)):

$$\phi_0 = \frac{{}_1V_1 - {}_2V_1}{S_0(b_1 - a_1)} = \frac{\frac{75}{2} - 10}{30(1 - 0)} = \frac{55}{60} = \frac{11}{12},$$

$$\psi_0 = e^{-r\Delta t}\left[{}_1V_1 - \frac{{}_1V_1 - {}_2V_1}{b_1 - a_1}(1 + b_1)\right] = \frac{3}{4}\left[\frac{75}{2} - \frac{55}{2} \times 2\right] = -\frac{105}{8}.$$

The hedging portfolio at $t = 0$ consists of $\dfrac{11}{12}$ units of stock and borrowing of $\pounds\dfrac{105}{8}$.

(2.c) Because we assume that $S_1 = {}_2S_1 = \pounds 30$ is realised, we need to find $({}_2\phi_1, {}_2\psi_1)$. We have

$$_2\phi_1 = \frac{{}_3f_2 - {}_4f_2}{{}_2S_1({}_2b_2 - {}_2a_2)} = \frac{20 - 0}{30} = \frac{2}{3},$$

$$_2\psi_1 = {}_2V_1 - {}_2S_1 \, {}_2\phi_1 = 10 - 30 \times \frac{2}{3} = -10.$$

At $t = 1$ we use the portfolio $(\phi_0, \psi_0) = \left(\dfrac{11}{12}, -\dfrac{105}{8}\right)$ of value ${}_2V_1 = \pounds 10$ when $S_1 = \pounds 30$ to sell $\dfrac{1}{4}$ unit of stock for £30 per unit and to repay $\pounds\dfrac{15}{2}$ to the bank in order to form the required new portfolio $({}_2\phi_1, \, {}_2\psi_1) = \left(\dfrac{2}{3}, -10\right)$.

Example 10.2. Consider Example 10.1. Assume that the real probability P for the price model in this example is such that each price movement at each step happens with probability $1/2$. Compute the Radon–Nikodym derivative Z of the risk-neutral measure Q from this example with respect to the real measure P. For your verification, the answer is

$$Z(`UU') = Z(`UD') = \frac{2}{3}, \quad Z(`DU') = \frac{16}{9}, \quad Z(`DD') = \frac{8}{9}.$$

10.3   Remark about 'Fair Price'

One of the important concepts in Financial Mathematics is the notion of 'fair price', which was mentioned in earlier chapters.

Definition 10.1. *The **fair price** of an option is the minimal initial wealth such that this wealth can be used so that the option obligation is fulfilled for any market situation via an $\mathcal{A}_n$-adapted strategy.*

The price x from (10.7) is the fair price for the European-type options in our market model because, first, it is enough to meet the obligation and, second, a smaller initial wealth is not sufficient to meet the obligation (see Section 9.5).

Chapter 11

Complete and Incomplete Markets

So far, we have always been able to perfectly hedge a contingent claim, in other words, we have considered only attainable claims. Due to Theorem 10.1, in the considered binary model (I) any European-style derivative claim is attainable (see Definition 9.9) by trading with a self-financing portfolio. Is this always so?

Example 11.1. Return to Example 7.1, where we dealt with the one-period binary model. We assumed that we knew that the asset price would be one of just two special values at time T. What if we allow *three* values at time T (see Fig. 11.1)?

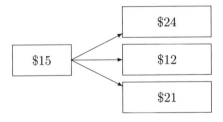

Fig. 11.1 One-period trinomial model.

Again, the seller would like to replicate the European call with the strike price $K = \$18$ and maturity T by a portfolio consisting of ϕ_0 stock and ψ_0 of money on the bank account. Assume that the interest rate is $r = 0$.

This time there will be three scenarios to consider, corresponding to the three possible values of S_T. This gives the three relations:

$$24\phi_0 + \psi_0 = 6 \qquad (11.1)$$
$$21\phi_0 + \psi_0 = 3$$
$$12\phi_0 + \psi_0 = 0.$$

What is the answer here? Well, it is obvious that we cannot find (ϕ_0, ψ_0) which satisfies the system of linear equations (11.1). If (ϕ_0, ψ_0) is such that at least one of the equalities becomes the inequality with the left-hand side being less than the right-hand side, the writer does not meet an obligation with positive probability. Let G be the region of the plane (ϕ_0, ψ_0) such that

$$24\phi_0 + \psi_0 \geq 6$$
$$21\phi_0 + \psi_0 \geq 3$$
$$12\phi_0 + \psi_0 \geq 0.$$

As has already been pointed out, there is no such point in G for which all the equalities in (11.1) are satisfied. Therefore, if (ϕ_0, ψ_0) belongs to G then there is an arbitrage opportunity: at any point $(\phi_0, \psi_0) \in G$, the seller has a strictly positive probability of making a profit and zero probability of making a loss. At the same time, any portfolio $(\phi_0, \psi_0) \notin G$ carries a risk of a loss. There is no portfolio that *exactly* replicates the claim. This market is not *complete*, i.e. there are contingent claims that cannot be perfectly hedged.

Definition 11.1. *A **complete market** is a market in which every contingent claim with maturity at a finite $T > 0$ and with bounded discounted payoff is attainable, otherwise the market is called **incomplete**.*

Remark 11.1. (*About bigger models*). In our market model we only have a single stock and a bank account. Real markets are more complex. We could modify the market in Example 11.1 by: (i) adding a third 'independent' asset and (ii) introducing an additional time moment between 0 and T on the tree. If we were to do this appropriately (see, e.g., Pliska, 1997), we obtain a complete market again. At the same time, a market model can be incomplete by its nature and the proposed adjustments might not work.

To get more intuition for the notion of complete/incomplete markets, we consider further examples.

Example 11.2. (*When a single-period market is complete*). Assume that a single-period market consists of M tradable assets in which prices follow a model such that at the end of the time period the market is in one of L

possible states. This market is complete if and only if the rank of the matrix of asset prices (which we denote by D) is L.

Answer. Any claim in this single-period market can be expressed as a vector $\mathbf{f} \in \mathbf{R}^L$. A replication of this claim is a portfolio $\vartheta = \vartheta(\mathbf{f}) \in \mathbf{R}^M$ for which $D\vartheta = \mathbf{f}$. To find such a portfolio, we need to solve the system of L linear equations in M unknowns. Thus (due to Linear Algebra[1]), a replicating portfolio exists for every possible realisation of $\mathbf{f}$ if and only if the rank of D is L. Note that in Example 11.1 we had $M = 2$, $L = 3$ and the rank of D was 2, which is less than the L.

Example 11.3. In a complete arbitrage-free market the risk-neutral measure Q is unique on $\mathcal{A}_N$.

Answer. Take any event $A \in \mathcal{A}_N$ and a claim $f := e^{rT} I_A$. Because the market is assumed to be complete, this claim is replicable with some initial wealth x. Further, there is an EMM Q because the market is assumed to be arbitrage free (see the first fundamental theorem of asset pricing). Then, thanks to Proposition 9.3 from Section 9.5, this initial wealth x is uniquely defined by $x = e^{-rT} E_Q f = E_Q I_A$ for any risk-neutral measure Q. Because $E_Q I_A = Q(A)$ and A was an arbitrary event from $\mathcal{A}_N$, Q is indeed uniquely defined on $\mathcal{A}_N$.

The statement opposite to the one in Example 11.3 is also true, although its proof[2] is skipped here. Summarising, the following result holds.

Second fundamental theorem of asset pricing. *An arbitrage-free market is complete if and only if there is a unique EMM.*

We observe that we can have the following four cases of market models:

1. Arbitrage-free complete market $\Leftrightarrow$ unique EMM (e.g., Example 11.2 and also the binary tree model (I)).
2. Arbitrage-free incomplete market $\Leftrightarrow$ non-unique EMM (see Example 11.5 below).

---

[1] Due to the compatibility criterion for a system of linear equations (also known as Kronecker–Capelli or Rouché–Capelli theorem), the linear system $D\vartheta = \mathbf{f}$ has a solution if and only if the rank of the coefficient matrix D is equal to the rank of the augmented matrix $[D|\mathbf{f}]$ obtained by appending the column of free terms $\mathbf{f}$ to the right of D. Because we require that $D\vartheta = \mathbf{f}$ has a solution for **any** $\mathbf{f}$, we should require that $M \geq L$ (i.e. that the number of unknowns is not less than the number of equations) and the rows of D are linearly independent, hence the rank of D is equal to L.

[2] It can be found, e.g., in Bingham and Kiesel (2004); Björk (2004); Melnikov (2004) (see also Harrison and Pliska (1981)).

3. Not free of arbitrage complete market – there is no EMM but every claim is hedgeable (see Example 11.4 below).
4. Not free of arbitrage incomplete market – there is no EMM and not every claim is hedgeable.

Example 11.4. (*Not free of arbitrage complete market*[3]). Let $\mathcal{A} = \{\Omega, \varnothing\}$ be the trivial algebra. Consider the continuous-time market consisting of a bank account with zero interest rate $B_t \equiv 1$ and an asset with the value $S_t = 100 + t$. This market model is deterministic. Show that this model is complete but not free of arbitrage.

Answer. First, show arbitrage existence in this model via the following strategy. At time $t = 0$ borrow 100 units of money from a bank and buy one asset for $S_0 = 100$. The initial value of this portfolio $V_0 = 0$. At any time $t > 0$ sell the one asset for $S_t = 100 + t$ and return 100 to the bank. The value of the portfolio at time t is $V_t = 100 + t - 100 = t > 0$ with certainty. Hence, this market is not free from arbitrage. Now show completeness. Consider an arbitrary function of time f_t so that f_T represents a claim at time T. Note that the market is deterministic, so it has only a single state at time T and f_T is a particular number. We have two instruments, the bank account and the asset. Suppose we form a portfolio at time $t = 0$ of ϕ_0 assets and ψ_0 of money in the bank account, which we can re-arrange at every moment t. In order to prove completeness of the market, we need to find such a self-financing strategy (ϕ_t, ψ_t), $0 \le t \le T$, that $V_T = \phi_T S_T + \psi_T$ is equal to f_T, i.e.,

$$\phi_T(100 + T) + \psi_T = f_T \tag{11.2}$$

and, e.g., the strategy $\phi_t = 0$, $\psi_t = f_T$ satisfies (11.2) and, hence the market is complete – for every f_T we can find a replicating self-financing strategy. Note that there are infinitely many strategies which satisfy (11.2).

Example 11.5. (*Continuation of Example 11.1*). We have already seen that the market model from Example 11.1 is incomplete. Is it arbitrage free? Due to Proposition 9.2, it is sufficient to show that there is EMM in this market, i.e. there is a measure Q which makes the discounted price process a martingale. In the case of Example 11.1, we need to find a Q such that (remember that the interest rate is zero in this example):

$$E_Q S_1 = S_0$$

---

[3]See in Filipovic (2009, p. 77).

or, equivalently,

$$24q_1 + 21q_2 + 12(1 - q_1 - q_2) = 15 \qquad (11.3)$$

with q_1, q_2 and $1 - q_1 - q_2$ satisfying the probability axioms and ensuring that all the three outcomes are possible under Q, i.e. q_1, q_2 and $1 - q_1 - q_2$ must be between zero and one. Simplifying (11.3), we get

$$4q_1 + 3q_2 = 1$$

and we can, e.g., choose $q_1 = 1/8$ and $q_2 = 1/6$ and form the EMM as

$$Q(S_1 = 24) = 1/8, \quad Q(S_1 = 21) = 1/6, \quad Q(S_1 = 12) = 17/24.$$

We note that this is one of infinitely many EMMs we can find for this market model. The existence of infinitely many EMMs implies that there can be infinitely many arbitrage prices for a particular claim in this market. Note that our conclusion here does not contradict Proposition 9.3 since there is no replicating strategy for the call on the market model from Example 11.1.

Remark 11.2. In this chapter we considered examples of markets which are incomplete because of a mismatch between the number of traded assets and the number of possible future states of the market. Incomplete markets also appear in other situations, for instance in markets with constraints (e.g., constraints on short selling) and friction (transaction costs). In general, complete markets are an idealistic model of real markets which, in many cases, can be considered as a good approximation for reality while incomplete markets are considered to be closer to reality.

Remark 11.3. In this course we are dealing only with market models which are arbitrage-free and complete, except for the current chapter. If the arbitrage-free assumption is rather natural from the real applications point of view and widely used in Finance, the completeness assumption is merely an approximation of reality because real markets are generically incomplete for various reasons (see Remark 11.2). We observed that in an incomplete market, if a claim f is not attainable, different martingale measures can give different prices. Then the natural question is how to choose the price with which both the buyer and seller might agree. The most commonly accepted ways of treating incomplete markets are via the use of utility functions or of strategies with consumption (see, e.g., Föllmer and Schied, 2004; Melnikov, 2004; Pliska, 1997), which we do not consider in this course.

Chapter 12

American Options

The most popular type of options on financial markets are American options. Although the labels (European and American options) apparently have some geographical justification, both are now traded everywhere in the world. Let us consider plain vanilla American options.

Definition 12.1. An *American call (put) option* with strike price K and expiry time T gives the holder the right, but not the obligation, to buy (sell) an asset for price K, at any time up to T.

Unlike a European derivative security, which can only be exercised at one time (expiration date), an American option entitles its owner to exercise it at any time prior to or at the expiration date.

In order to price an American option in the complete market setting, as we did in the case of European options, we use the idea of replication: we imagine that the writer sells the option for an initial capital (premium) and then uses this initial capital to hedge. In the case of an American option, the writer needs to have funds to meet his obligation at all times before and at the maturity because he does not know when the option will be exercised. To price this type of option:

1. The writer has to determine the worst (from his perspective) time at which the holder can exercise the option. For the holders, this is the **optimal exercise time**.
2. The writer then computes the premium which he needs to replicate the claim at the optimal exercise time.

With the pricing problem solved, the writer then answers the hedging question, i.e. how to invest the premium so that the writer can meet his obligation even if the holder exercises at a non-optimal time.

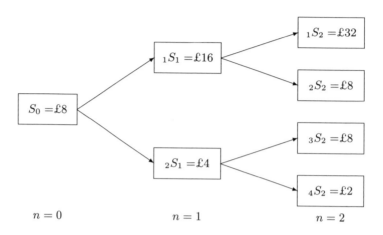

Fig. 12.1   The tree diagram for Example 12.1.

Let us compare American and European options. Obviously, the holder may exercise the American option at the maturity time T, as in the case of its European counterpart. In order to hedge such a claim, the writer should have the same initial capital (premium) as in the 'European' case. Hence, the value of an American option should be at least not less than that of the corresponding European one. The possibility to get a payoff at an earlier time tends to make American options more expensive than their European counterparts: the better (and here better because more flexible) the product the higher the price is. Surprisingly, however, there are important situations when values of American and the corresponding European options coincide. We will consider these a bit later on but first an example.

Example 12.1. Consider the two-period model for the stock price from Fig. 12.1. Let the continuously compounded interest rate r be such that $e^{r\Delta t} = 5/4$. Find the value of an American put option with maturity $T = 2\Delta t$ and strike $K = 9$.

Answer. The considered derivative is such that the holder of the option can exercise it at any time n between 0 and 2 to receive $f(S_n) = (9 - S_n)_+$.

Let us first make some preparations. In this model the parameters of (I) are

$$a_n = a = -\frac{1}{2} \text{ and } b_n = b = 1,$$

i.e. it is a binomial tree. We note that $1 + a = \dfrac{1}{2} < e^{r\Delta t} = \dfrac{5}{4} < 1 + b = 2$,
i.e. the no-arbitrage condition (Ic) is satisfied. We also see that all the prices
are positive, i.e. the positiveness condition is satisfied as well. So, the model
of this example is consistent with our conditions for model (I).

As in the case of European options, let us find the risk-free probabilities.
Recall

$$q = \frac{e^{r\Delta t} - (1 + a)}{b - a}.$$

Then, for this model we have

$$q = \frac{\frac{5}{4} - (1 - \frac{1}{2})}{1 + \frac{1}{2}} = \frac{1}{2} \text{ and } 1 - q = \frac{1}{2},$$

i.e. in this example the risk-free probability of stepping up (down) is $1/2$ at
every node.

Now we are ready to find the value of this American put option. The
value of the claim at time $n = 2$ is

$$_1V_2 := f(_1S_2) = 0,$$
$$_2V_2 := f(_2S_2) = 1, \quad _3V_2 := f(_3S_2) = 1,$$
$$_4V_2 := f(_4S_2) = 7.$$

At time $n = 1$, we should consider two possibilities:

1. The value of the portfolio if the holder does not exercise the claim at
   $n = 1$ (then the writer should be able to meet the obligation at time
   $n = 2$).
2. The value of the portfolio if the holder does exercise the claim at $n = 1$.

Consider the first case in the same way as we did in the case of European
options. To meet the obligation, the writer should have a portfolio with the
value U_1 at $n = 1$, such that

$$U_1 = e^{-r\Delta t} E_Q(f(S_2)|\mathcal{A}_1) = e^{-r\Delta t} E_Q((5 - S_2)_+|\mathcal{A}_1).$$

Hence

$$\text{for first path: } _1U_1 = \frac{4}{5}(\frac{1}{2} \times 0 + \frac{1}{2} \times 1) = \frac{2}{5},$$
$$\text{for second path: } _2U_1 = \frac{4}{5}(\frac{1}{2} \times 1 + \frac{1}{2} \times 7) = \frac{16}{5}.$$

Now the second case. If the holder exercises at $n = 1$, the writer should have a portfolio with the value W_1 so that

$$\text{for first path: } {}_1W_1 = f({}_1S_1) = 0,$$
$$\text{for second path: } {}_2W_1 = f({}_2S_1) = 5.$$

Combine these two cases. The writer can hedge this option if at time $n = 1$ his wealth V_1 is

$$V_1 = \max(W_1, U_1),$$

i.e.

$$\text{for first path: } {}_1V_1 = \max({}_1W_1,\ {}_1U_1) = \max(0, \frac{2}{5}) = \frac{2}{5},$$
$$\text{for second path: } {}_2V_1 = \max({}_2W_1,\ {}_2U_1) = \max(\frac{16}{5}, 5) = 5.$$

If the portfolio has a smaller value than V_1 then the writer will not be able to meet his obligation in one or another case (and here you see the difference with European options).

Now we move onto $n = 0$. At $n = 0$, the writer should have a portfolio such that it will be enough to meet the obligations ${}_1V_1$ and ${}_2V_1$ at $n = 1$, which are known to us at time $n = 0$. As in the case of European options, we get

$$U_0 = e^{-r\Delta t} E_Q V_1 = \frac{4}{5} \times \frac{1}{2} \left(\frac{2}{5} + 5 \right) = \frac{54}{25}.$$

In addition, the option can be exercised at $n = 0$ requiring the writer to have at least the wealth $W_0 = 1$. Then, the premium (value of the American option) is:

$$x = V_0 = \max(W_0, U_0) = \max\left(1, \frac{34}{25} \right) = \frac{54}{25}\ (\pounds).$$

12.1 Stopping Times

The *optimal stopping time* τ^* (in our situation it is the optimal exercise time – the time at which an American derivative should be exercised to get maximum possible profit by the rational holder) is a random variable; it depends on the underlier's price trajectory. In other words, on each path of the underlier it might be different. We start with a motivation example.

Example 12.2. Let us look back at Example 12.1. What is the best time (i.e. the optimal exercise time) for the holder to exercise the option? First, he does not exercise the option at $n = 0$ because the payoff $f(8) = £1$, which he would get, is less than the price of the option $x = £54/25$. Let us consider the next steps.

U* At time $n = 1$, if the price goes up, the holder gets nothing for the put (the put is OTM) and he has nothing to exercise.

•UU If at time $n = 2$ the price is further up, the put is again OTM and the holder again has nothing to exercise. In this case the random variable $\tau^* = \tau^*(\omega)$ – the optimal stopping time – takes the value ∞ ('dead' state), by which we mean that the option should be allowed to expire without exercise and we write $\tau^*('UU') = \infty$.

•UD Consider $n = 2$ and the path $\omega = 'UD'$, when $S_2 = {}_2S_2 = 8$. Here the put is ITM and the holder should exercise the option, and we write $\tau^*('UD') = 2$.

D* What does the holder do at time $n = 1$ if the price went down? The put is ITM. Here, this is a trickier question: to exercise or not? The holder can get £5 if he exercises at time $n = 1$. We note that next period, at $n = 2$, the option will be worth either $f(8) = £1$ or $f(2) = £7$. The risk-neutral pricing formula told us (see Example 12.1) that to construct a hedge against these two possibilities, at time $n = 1$ we had to have a portfolio with the value ${}_2V_1 = £16/5 = £3.2$. So, if we have £3.2 at $n = 1$, we can receive either £1 or £7 depending on the price movement. Then, it is unwise for the rational holder not to exercise the option: he gets the payoff £5, can consume £1.8 and continue the hedge with the remaining £3.2 value in the portfolio. In other words, if the holder refuses to exercise, the value of the option becomes £3.2, while by exercising he gets £5. Thus, in this case the optimal stopping time is $n = 1$ and we write $\tau^*('DU') = \tau^*('DD') = 1$.

In this example the random variable τ^* is defined on $\Omega = \{'UU', 'UD', 'DU', 'DD'\}$ and it takes values in the set $\{0, 1, 2, \infty\}$, which can be represented in Table 12.1.

Table 12.1 The optimal stopping time in Example 12.2.

ω	'UU'	'UD'	'DU'	'DD'
τ^*	∞	2	1	1

You might have noticed that the holder of the put in the above example will regret that he did not exercise the option at time zero if the first price movement was up (earlier is better than later). Further, if the holder had the *a priori* information of the price behaviour, he would be better off using the following exercise rule than the one we found in Example 12.2:

$$\rho(\text{'}UU\text{'}) = 0, \quad \rho(\text{'}UD\text{'}) = 0, \quad \rho(\text{'}DU\text{'}) = 1, \quad \rho(\text{'}DD\text{'}) = 2. \qquad (12.1)$$

Using this exercise rule (12.1), the holder would always exercise the put in the money regardless of price movements. However, there is a problem with this exercise rule ρ: it cannot be used without prior information (i.e. without knowing the future), which we can call 'insider information'. Indeed, this rule is such that, e.g., the decision of whether or not to exercise at time zero is based on the outcome of the first price movement. So, this rule ρ contradicts what common sense suggests we are able to do.

We now introduce the notion of stopping time rigorously.

Definition 12.2. *(Discrete time). A random variable* $\tau = \tau(\omega)$ *taking values in the set* $\{0, 1, 2, \ldots, N, \infty\}$ *is called a **stopping time** or a **Markov time** with respect to algebras* $\mathcal{A}_n$ *if the event*

$$\{\omega : \tau(\omega) = n\} \in \mathcal{A}_n \qquad (12.2)$$

for any $n = 0, \ldots, N$; *and if* $\tau(\omega) \neq n$, $n = 0, \ldots, N$, *we set* $\tau(\omega) = \infty$.

Note that the event $\{\tau(\omega) \neq n, \ n = 0, \ldots, N\} \in \mathcal{A}_N$.

This definition means that a time τ is a stopping time if, having the information $\mathcal{A}_n$, we can say at time n whether the event $\tau(\omega) = n$ happens or not. Then the stopping time is also called a random variable independent of the 'future'. Roughly speaking, a stopping time is a random time that makes the decision to stop (e.g., exercise an option) without looking ahead (see also the discussion below, after (12.3)).

Example 12.3. It is clear that the time ρ from (12.1) is not a stopping time whereas the optimal stopping time τ^* from Example 12.2 satisfies the condition (12.2) and it is a stopping time. Let us check this formally. In Example 12.2 we have $\Omega = \{\text{'}UU\text{'}, \text{'}UD\text{'}, \text{'}DU\text{'}, \text{'}DD\text{'}\}$, $\mathcal{A}_0 = \{\varnothing, \Omega\}$, $\mathcal{A}_1 = \{\varnothing, \Omega, A, \bar{A}\}$ with $A = \{\text{'}DU\text{'}, \text{'}DD\text{'}\}$, and $\mathcal{A}_2 = \mathcal{A}$, which consists of all the subsets of Ω including $\varnothing$. The time ρ from (12.1) is not a stopping time because, e.g., $\bar{A} \notin \mathcal{A}_0$ but $\{\omega : \rho(\omega) = 0\} = \bar{A} \in \mathcal{A}_1$. For τ^* from Example 12.2, we have $\{\omega : \tau^*(\omega) = 1\} = A \in \mathcal{A}_1$, $\tau^*(\text{'}UD\text{'}) = 2$ and $\text{'}UD\text{'} \in \mathcal{A}_2$, $\tau^*(\text{'}UU\text{'}) = \infty$ and $\text{'}UU\text{'} \in \mathcal{A}_2$; hence this τ^* is indeed a stopping time.

Example 12.4. Another example: the constant function $\tau(\omega) \equiv n$ is a stopping time for a fixed n.

One can prove that τ is a stopping time if and only if

$$\{\omega : \tau(\omega) \le n\} \in \mathcal{A}_n \qquad (12.3)$$

for any $n \ge 0$. The random moment $\tau(\omega)$ is the instant when one must make a certain decision to take an action like to exercise an option, buy a stock or stop gambling, etc. The condition (12.3) is equivalent to $\{\omega : \tau(\omega) > n\} \in \mathcal{A}_n$ (see Definition 8.3 of an algebra), which can be interpreted as follows. If we decide at time n to postpone a certain action, the only information we could use for making this decision is $\mathcal{A}_n$, i.e. the information available to us up to time n only, as we cannot take into account the 'future', i.e. what might happen after time n (because we do not know it yet!). In a sense the concept of stopping time explains why we cannot catch the best time to, e.g., sell a stock at the maximal price over a given time interval $[0, T]$ or to stop gambling: this time, of course, exists in $[0, T]$ but we can only determine it by observing data over the whole period $[0, T]$ and so we can find this best time only *a posteriori*. Note that the notion of stopping time has no connection with probability P on Ω, it is related only to the sample space Ω and algebras in it.

We can also reformulate Definition 12.2 for our purposes motivated by pricing and hedging American options in the discrete-time setting as follows. We say that in an N-period binary model, the stopping time is a random variable τ that takes values $0, 1, \ldots, N$ or ∞ and satisfies the condition that, if for a particular path $(c_1, c_2, \ldots, c_n, c_{n+1}, \ldots, c_N)$ of the N-period binary model the stopping time is equal to n:

$$\tau(c_1, c_2, \ldots, c_n, c_{n+1}, \ldots, c_N) = n,$$

then it is also equal to n for all other paths in which the portions $(c_1, c_2, \ldots, c_n)$ from 0 to n are the same, i.e.

$$\tau(c_1, c_2, \ldots, c_n, c'_{n+1}, \ldots, c'_N) = n \quad \text{for all } c'_{n+1}, \ldots, c'_N,$$

where $c_i, c'_i \in \{\text{'U'}, \text{'D'}\}$. As before, we set $\tau = \infty$ if $\tau(c_1, \ldots, c_N) \ne n$, $n = 0, \ldots, N$.

One can show that if τ and θ are two stopping times then the following functions are also stopping times[1]:

$$\tau \wedge \theta, \quad \tau \vee \theta, \quad (\tau + \theta) \wedge N.$$

---

[1]Here $a \wedge b := \min(a, b)$, $a \vee b := \max(a, b)$.

Having in mind our N-period binary model, we also introduce a sequence of stopping times τ_k with respect to algebra $\mathcal{A}_n$ with values in $\{k, \ldots, N, \infty\}$, where $k = 0, 1, \ldots, N$, i.e.

$$\{\omega : \tau_k(\omega) \leq n\} \in \mathcal{A}_n \quad \text{for } n \geq k.$$

For instance, if we forgot about an American option we possess until the time k and starting from this time k we would like to maximise our profit then the optimal exercise time τ_k^* for the American option in the time period $n \geq k$ is an example of stopping times τ_k.

Let $\mathbb{M}_k$, $k = 0, \ldots, N$, be the set of all possible stopping times τ_k, i.e. all possible random variables with values in $\{k, \ldots, N, \infty\}$ defined on Ω and independent of the 'future'. In particular, the set $\mathbb{M}_0$ contains every stopping time. A stopping time τ_N in $\mathbb{M}_N$ can take the value N on some paths and the value ∞ on others; it cannot take any other value.

12.2 Pricing and Hedging American Options on the Binary Tree

Let us now formalise pricing of American options in the discrete case. As before, we assume that the stock price is due to (I) and we have a payoff function $f(s)$ (e.g., in the case of an American call option $f(s) = (s - K)_+$) and a maturity time $T = N\Delta t$. At any time $0 \leq n \leq N$, the holder of an American option can exercise and receive payment $f(S_n)$, which is called the *intrinsic value* of the option. Thus, the portfolio of the writer should always have the value V_n satisfying

$$V_n \geq f(S_n), \; n = 0, 1, \ldots, N, \tag{12.4}$$

(obviously for any possible realisation of S_n), i.e. the value of an American derivative security is always greater or equal to its intrinsic value. This suggests (together with Example 12.1) that to price an American derivative security, we need a new pricing algorithm, the American option pricing algorithm. In this algorithm (which is, as usual, backwards in time) we evaluate the price of an American option and find the stopping time.

As in the 'European' case, we introduce the portfolio sequence (ϕ_n, ψ_n), $n = 0, \ldots, N - 1$, where ϕ_n is the amount of stock held by the writer during the $(n + 1)$ period and ψ_n is the amount of money deposited by the writer on the bank account at time n, and we consider strategies (ϕ_n, ψ_n), which are $\mathcal{A}_n$-adaptive and self-financing. Further, let $\tau_k^* \in \mathbb{M}_k$ be the optimal exercise time for the American option in the time period $n \geq k$.

We start constructing the backward algorithm by observing that at the maturity time N the writer has the obligation $f(S_N)$ and the value of his replicating portfolio formed at time $N-1$ should match this obligation:

$$V_N = S_N \phi_{N-1} + e^{r\Delta t} \psi_{N-1} = f(S_N). \tag{12.5}$$

The stopping time $\tau_N^* \in \mathbb{M}_N$ can only take the two values:

$$\tau_N^* = N \ \text{ if } \ f(S_N) > 0 \ \text{ and } \ \tau_N^* = \infty \ \text{ otherwise.}$$

At the next time, $n = N-1$, to hedge the future obligation (12.5) the writer's wealth should be $V_{N-1} \geq U_{N-1}$, where U_{N-1} is the funds needed to buy the portfolio (ϕ_{N-1}, ψ_{N-1}) to satisfy (12.5) and it is found analogously to the 'European' case (see Section 10.1). Indeed, in each 'block' of the binary tree in the period $N-1$ we find the portfolio which meets the obligation at $n = N$:

$$\phi_{N-1}\, {}_U S_N + e^{r\Delta t} \psi_{N-1} = {}_U V_N$$

$$\phi_{N-1}\, {}_D S_N + e^{r\Delta t} \psi_{N-1} = {}_D V_N$$

(the notation here is analogous to the one used in Section 10.1). Then

$$U_{N-1} = \phi_{N-1}\, S_{N-1} + \psi_{N-1},$$

which we can rewrite (as usual) in the form of conditional expectation

$$U_{N-1} = e^{-r\Delta t} E_Q[f(S_N)|\mathcal{A}_{N-1}],$$

where Q is the risk-free measure from (II). The U_{N-1} is the value of holding rather than exercising the option at time $n = N-1$ and it is called the *continuation value*. At the same time, due to (12.4), the writer should have a portfolio at time $n = N-1$ with the value $V_{N-1} \geq f(S_{N-1})$, and, consequently, we require that

$$V_{N-1} = \max\{f(S_{N-1}), U_{N-1}\}.$$

By the same arguments as in the previous section (see Example 12.2), if $f(S_{N-1}) \geq U_{N-1}$ and $f(S_{N-1}) > 0$, the holder should exercise, otherwise he should postpone his decision until a later step (the step N here). Then

$$\tau_{N-1}^* = \begin{cases} N-1 \text{ if } f(S_{N-1}) \geq U_{N-1} \text{ and } f(S_{N-1}) > 0; \\ \tau_N^* \qquad \text{otherwise.} \end{cases}$$

The value V_{N-1} is the obligation translated to the period $n = N-2$. Then we analogously get at $n = N-2$:

$$U_{N-2} = e^{-r\Delta t} E_Q[V_{N-1}|\mathcal{A}_{N-2}],$$

$$V_{N-2} = \max\{f(S_{N-2}), U_{N-2}\},$$

$$\tau_{N-2}^* = \begin{cases} N-2 \text{ if } f(S_{N-2}) \geq U_{N-2} \text{ and } f(S_{N-2}) > 0; \\ \tau_{N-1}^* \qquad \text{otherwise.} \end{cases}$$

Continuing by induction, we obtain the **American option pricing algorithm**[2]:

$$V_N = f(S_N); \tag{12.6}$$

$$V_k = \max\{f(S_k), U_k\}, \quad U_k = e^{-r\Delta t} E_Q[V_{k+1}|\mathcal{A}_k],$$

$$\tau_k^* = \inf_{k \leq m \leq N}\{m : V_m = f(S_m) \text{ and } f(S_m) > 0\},$$

$$k = N - 1, \ldots, 0.$$

Here, U_k is the continuation value at time k and V_k gives the option's price at time $n = k$. In particular, at time $n = 0$ the price of American option is

$$x = V_0 = \max\{f(S_0), e^{-r\Delta t} E_Q V_1\}$$

and the optimal exercise time is $\tau^* = \tau_0^*$. Let us emphasise again that τ^* is a random variable, it depends on a trajectory of the price process S_n.

Remark 12.1. The pricing algorithm for American-type options with path-dependent payoffs (e.g., American-style Asian options, i.e. Asian options with the early exercise feature – see Example 9.11, where we allow options to be exercised at any time from 0 to N) is essentially the same as the algorithm (12.6).

The algorithm (12.6) shows that we can replicate any American claim $f(s)$ in the case of model (I). Now we rewrite this algorithm in the forward time. For obvious reasons, we give it until the optimal exercise time. Due to the American option pricing algorithm, it is evident that the hedging algorithm in the 'American' case is the same as the 'European' one until the stopping time.

HSA: the hedging strategy (algorithm for the writer)

1. Calculate the premium $x = V_0$ according to (12.6) and find the optimal stopping time τ^*.
2. Set $n = 0$ (start the 'clock') and charge the premium $x = V_0$.
3. If $\tau^* = n$ then the holder exercises the option, the writer has the portfolio with the value V_{τ^*} which matches the payoff function $f(S_{\tau^*})$, and the algorithm stops; otherwise go to 4.
4. (a) Evaluate the delta

$$\Delta_{n+1} = \frac{_U V_{n+1} - _D V_{n+1}}{S_n(b_{n+1} - a_{n+1})},$$

where $_U V_{n+1}, _D V_{n+1}$ are computed according to (12.6);

---

[2]Recall that $\inf(\mathcal{S})$ is the biggest real number that is less than or equal to every member of $\mathcal{S}$ and if $\mathcal{S}$ is empty, $\inf(\mathcal{S}) = \infty$.

(b) Using the existing funds V_n:
- buy $\phi_n = \Delta_{n+1}$ units of stock (S_n each),
- put $\psi_n = V_n - \phi_n S_n$ on the bank account.

5. (a) Arrive at next time, $n + 1$, and get value of S_{n+1} from the market;
   (b) Revaluate the wealth: $V_{n+1} = \phi_n S_{n+1} + e^{r\Delta t}\psi_n$;
   (c) Redefine $n := n + 1$.

6. If $n = N$ and $\tau^* = \infty$, then the algorithm stops – the option expires without exercise; otherwise go to 3.

As we see from the algorithm HSA, before the exercise time τ^*, we hedge the American option in the same way as the European one and at the exercise time the writer should be ready to pay $f(S_{\tau^*})$ to the holder. Consequently, if we know the *optimal exercise boundary* $S_{\tau^*}(\omega)$ (i.e. values of the stock price $_iS_n$ such that a trajectory satisfies the following conditions for the first time: $V_n(_iS_n) = f(_iS_n)$ and $f(_iS_n) > 0$) then the price of the American option is given by

$$x = V_0 = E_Q\left[e^{-r\tau^*\Delta t}f(S_{\tau^*})I_{\{\tau^* \leq N\}}\right]. \tag{12.7}$$

The term $I_{\{\tau^* \leq N\}}$ appears in (12.7) because the value under the expectation should be replaced by zero on those paths for which $\tau^* = \infty$. Note that τ^* is a random variable and functions of it cannot be taken out of the expectation in (12.7).

The optimal exercise boundary $S_{\tau^*}(\omega)$ is the boundary between the so-called *stopping domain* (i.e. a time-price domain $(n, {}_iS_n)$ such that if (n, S_n) belongs to it then it is optimal for the holder to exercise the option) and *continuation domain* (i.e. a domain $(n, {}_iS_n)$ such that if (n, S_n) belongs to it then it is optimal for the holder to continue to hold the option).

We also note that it is not difficult to see from the derivation of (12.6) and (12.7) that

$$V_k = E_Q\left[e^{-r(\tau_k^* - k)\Delta t}f(S_{\tau_k^*})I_{\{\tau_k^* \leq N\}}|\mathcal{A}_k\right]. \tag{12.8}$$

Further, in our derivation of (12.6) we choose a stopping time τ_k^* from the collection $\mathbb{M}_k$ of all possible stopping times τ_k at each step k such that

$$E_Q\left[e^{-r(\tau_k - k)\Delta t}f(S_{\tau_k})I_{\{\tau_k \leq N\}}|\mathcal{A}_k\right]$$

reaches its maximum (remember that we stop at the optimal time, the time at which the holder gets maximum possible payoff), i.e.

$$V_k = \max_{\tau_k \in \mathbb{M}_k} E_Q\left[e^{-r(\tau_k - k)\Delta t}f(S_{\tau_k})I_{\{\tau_k \leq N\}}|\mathcal{A}_k\right]. \tag{12.9}$$

In particular, we have[3]

$$x = V_0 = E_Q\left[e^{-r\tau^*\Delta t}f(S_{\tau^*})I_{\{\tau^* \leq N\}}\right] \tag{12.10}$$

$$= \max_{\tau_0 \in \mathbb{M}_0} E_Q\left[e^{-r\tau_0\Delta t}f(S_{\tau_0})I_{\{\tau_0 \leq N\}}\right].$$

Let us introduce another class of stochastic processes.

Definition 12.3. *(Discrete case). A process ζ_n is a **supermartingale** with respect to a measure Q and a filtration $\{\mathcal{A}_i\}_{i \geq 0}$ if*

$$E_Q|\zeta_m| < \infty, \quad E_Q(\zeta_{m+1}|\mathcal{A}_m) \leq \zeta_m \quad \text{for all } m. \tag{12.11}$$

Example 12.5. We know from Chapter 9 that in the case of an arbitrage-free complete market the discounted value of the portfolio replicating a European option is a martingale under the risk-neutral measure Q. Is this also true in the 'American' case?

Answer. Introduce the discounted wealth process

$$\tilde{V}_k := e^{-rk\Delta t}V_k.$$

It is usually called the *Snell envelope* of $\tilde{W}_k := e^{-rk\Delta t}f(S_k)$.

Due to (12.6), we have that in general:

$$\tilde{V}_k = \max\{\tilde{W}_k, E_Q[\tilde{V}_{k+1}|\mathcal{A}_k]\} \geq E_Q[\tilde{V}_{k+1}|\mathcal{A}_k],$$

i.e. the discounted wealth process is, in general, not a martingale under Q in the case of American options, it is a supermartingale under the risk-neutral measure. This observation together with properties of supermartingales is used for a more detailed study of American options (see, e.g., Föllmer and Schied, 2004; Shiryaev, 1999), which we do not consider in this course. Note also that in the next section we will see an example of when American and European options coincide and in that case the corresponding discounting wealth is a martingale, which is of course still a supermartingale.

Remark 12.2. In a sense, an intermediate option between an American and European option is *Bermudian*, which is an option that permits early exercise but only on a contractually specified finite set of dates before the maturity. The principle of pricing Bermudian options on the binary tree is effectively the same as for American ones.

Remark 12.3. If we change the inequality in (12.11) as $E_Q(\zeta_{m+1}|\mathcal{A}_m) \geq \zeta_m$ then the corresponding process is called a *submartingale* with respect to the measure Q.

---

[3]We note that though we demonstrated this result in the case of the binary tree price model, it is applicable more widely (see, e.g., Föllmer and Schied, 2004; Shiryaev, 1999).

12.3 When the Values of European and American Options Coincide

Let us now return to the point that there are important situations when the values of European and American options coincide, i.e. when there is no gain from an early exercise. Recall (see Section 10.1) that

$$U_n = e^{-r(N-n)\Delta t} E_Q[f(S_N)|\mathcal{A}_n]$$

gives us the fair price at time n of a European option with payoff $f(s)$ and maturity N. If for all n

$$f(S_n) \leq U_n \tag{12.12}$$

(i.e. when there is no gain from early exercise) then exercising the American option at time $n < N$ would be irrational because at time n we can always generate the payoff U_n. Indeed, we can follow, for instance, the following strategy.

1. At $t = n\Delta t$ – we have the American option that satisfies (12.12) and we sell the corresponding European option (i.e. for the same stock, with the same payoff, strike price and maturity time) for the fair price U_n.
2. At $T = N\Delta t$ – we exercise the American option and cover the obligation due to the European option if it is exercised.

Hence, by this scheme, we can receive the payment U_n at $t = n\Delta t$ if we have the American option, consequently, it is foolish to exercise it at $t = n\Delta t$ to get $f(S_n) \leq U_n$ (we would be losing money with certainty if $f(S_n)$ is strictly less than U_n). We can repeat this argument at each $n < N$. Thus, in the case of (12.12) it is optimal to wait until time N to decide whether to exercise. Therefore, the values of European and American options should be the same in this case. Note that (12.12) is not satisfied in Example 12.1 (i.e. (12.12) is not always true and we do need the American pricing algorithm (12.6)).

Now our aim is to discover for which functions f the inequality (12.12) holds. To this end, we recall the definition of convex functions.

Definition 12.4. *A function $f(s)$, $s \in \mathbf{R}$, is called a **convex** function if for any $s_1, s_2 \in \mathbf{R}$ and $\lambda \in [0, 1]$ we have*

$$f(\lambda s_1 + (1 - \lambda)s_2) \leq \lambda f(s_1) + (1 - \lambda)f(s_2).$$

To understand the definition, recall that $\lambda s_1 + (1 - \lambda)s_2$ is simply a line segment connecting s_1 and s_2 (in the s direction) and $\lambda f(s_1) + (1 - \lambda)f(s_2)$ is a line segment connecting $f(s_1)$ and $f(s_2)$. Pictorially, the function is convex in $\mathbf{R}$ if it lies below the straight line segment connecting any two points in $\mathbf{R}$ (see Fig. 12.2).

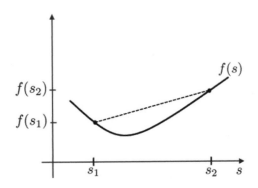

Fig. 12.2   Illustration of a convex function for Definition 12.4.

Example 12.6. Convex functions are: $f(s) = (s - K)_+$ – payoff of a call; $f(s) = (K - s)_+$ – payoff of a put with strike K. *Exercise.* Prove that this is true.

It can be proved that Definition 12.4 is equivalent to the following one.

Definition 12.5. *If f is **convex** on $\mathbf{R}$, for each $s_0 \in \mathbf{R}$ there is a number $\alpha(s_0)$ such that*

$$f(s) \geq f(s_0) + (s - s_0)\alpha(s_0) \text{ for all } s \in \mathbf{R}.$$

Now we state and prove an important inequality for expectations.

Jensen's inequality. *Let $f(s)$ be convex for each $s \in \mathbf{R}$ and ξ be a random variable with $E|\xi| < \infty$. Then*

$$f(E\xi) \leq Ef(\xi). \tag{12.13}$$

Proof. Putting $s = \xi$ and $s_0 = E\xi$ in Definition 12.5, we get

$$f(\xi) \geq f(E\xi) + (\xi - E\xi)\alpha(E\xi).$$

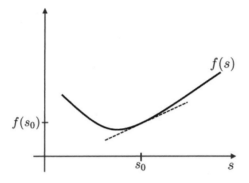

Fig. 12.3   Illustration of a convex function for Definition 12.5.

Taking expectation from both sides, we obtain

$$Ef(\xi) \geq E[f(E\xi) + (\xi - E\xi)\alpha(E\xi)] = f(E\xi),$$

which completes the proof of Jensen's inequality.

More generally, Jensen's inequality applies to conditional expectations, e.g.,

$$f(E(\xi|\mathcal{A}_n)) \leq E(f(\xi)|\mathcal{A}_n), \tag{12.14}$$

where $f(s)$ is a convex function for each $s \in \mathbf{R}$.

Now we limit our consideration to convex functions $f : [0, \infty) \to \mathbf{R}$ such that $f(0) = 0$. For instance, the payoff for a call: $f(s) = (s - K)_+$ satisfies both these requirements but the payoff of a put $f(0) = (K - 0)_+ \neq 0$. In Definition 12.4 of convex functions we choose $s_1 = s$ and $s_2 = 0$ and, taking into account that $f(0) = 0$, we get

$$f(\lambda s) \leq \lambda f(s) + (1 - \lambda)f(0) = \lambda f(s), \quad \lambda \in [0, 1]. \tag{12.15}$$

Recall (see Example 9.7) that the discounted stock price $e^{-rn\Delta t}S_n$ is a martingale under the risk-neutral measure then, using (9.8), we obtain

$$S_n = E_Q[e^{-r(N-n)\Delta t}S_N|\mathcal{A}_n]. \tag{12.16}$$

Therefore,

$$f(S_n) \overset{(12.16)}{=} f(E_Q[e^{-r(N-n)\Delta t}S_N|\mathcal{A}_n]) \overset{(12.14)}{\leq} E_Q[f(e^{-r(N-n)\Delta t}S_N)|\mathcal{A}_n]$$

$$\overset{(12.15)}{\leq} e^{-r(N-n)\Delta t}E_Q[f(S_N)|\mathcal{A}_n] = U_n,$$

i.e. we proved that (12.12) holds for convex functions $f : [0, \infty) \to \mathbf{R}$ with $f(0) = 0$. We summarise this result with the following theorem.

Theorem 12.1. *Consider the binary model (I). In this model, consider an American security with convex payoff function $f(s)$ satisfying $f(0) = 0$ and with maturity time $T = N\Delta t$. The value of this option is the same as the value of the corresponding European option.*

This theorem tells us that the early exercise feature of the American call does not contribute to its value: its value equals the value of the corresponding European call. The case of the American put is harder and we have to use the American option pricing algorithm (see Example 12.1, where we have considered an American put and seen that the early exercise feature does contribute to its value).

Remark 12.4. Note that we are considering a stock that does not pay dividends. If stock pays dividends then the prices of American and European calls might not coincide.

Exercise 12.1. Consider a European call option with value C_t and exercise price K on a stock with price S_t. Show by arbitrage arguments that $C_t \geq (S_t - K)_+$ for all $t \leq T$.

Example 12.7. It immediately follows from the statement of the above exercise that the payoff of a call satisfies (12.12), which provides another (actually easier) proof of why the prices of a European call and the corresponding American call coincide. We note that Theorem 12.1 is applicable more widely than just to calls.

Exercise 12.2. Complete the argument in Example 12.7: why Exercise 12.1 implies (12.12).

In the second part of the course we revised and extended our knowledge of Probability theory in the simplest setting of experiments with a finite number of outcomes. We also covered the basics of stochastic processes, in particular, we introduced information flow (filtration) and martingales. We used this knowledge for arbitrage pricing and hedging European and American options with underliers modelled by a binary tree and, in particular, obtained the discrete Black–Scholes formula. We were mainly dealing with complete market models, though we learned about their limitations, and briefly discussed incomplete markets. The central results of this part are two fundamental asset pricing theorems. The first one links the existence of an EMM and no arbitrage in a market model. The second fundamental theorem relates the uniqueness of EMM to a market being complete.

In the third and final part of the course we will introduce continuous-time modelling and will price and hedge European options in the case of underliers modelled by geometric Brownian motion. To this end, we will cover some basics of Ito Calculus. The main outcome of the third part will be the famous Black–Scholes formula which we will derive in two ways: first via a limit of the discrete Black–Scholes formula obtained in this second part of the course and then via continuous-time modelling.

Chapter 13

Problems for Part II

1. (*Probability revision*). Consider a three-period binomial model with the parameters $S_0 = 1$, $a = -0.5$, $b = 1$. Assume that the 'original' probability for the price to go up is always 1/3. Also, let us assume that price movements are independent.

(a) What is the sample space for this experiment? Assign probabilities to the elementary events.

(b) Define the random variable ξ = total number of 'up' movements.
   • Find its probability distribution.
   • Compute the expectation $E\xi$ and the variance $Var\xi$ (with respect to the 'original' probability).

(c) Find the probability that two 'up' movements of price happen assuming that there is at least one 'up' movement.

2. (*Probability revision*). Let us play a dice game. Each time you throw two fair dice and the game's outcome is as follows: if the sum of the dice is six then you win w, otherwise you lose. Find the smallest value of w that makes the game worth playing. (Assume that we will play for a while and throw the dice many times).

3. Let the three-step binary price model with the time step $\Delta t = 1/52$ (one week) have the structure shown in Fig. 13.1. Suppose the continuously compounded interest rate p.a. is such that $r = 52 \ln (5/4)$.

(i) Demonstrate that there are no arbitrage opportunities in this market model and find the risk-neutral probability for this model.

(ii) Consider the European put option with strike price equal to £16 and maturity time $n = 3$.

(a) Find its value at time $t = 0$;

(b) Suppose the writer follows a perfect hedging strategy. What portfolio should he have at time $n = 1$ if the stock price at this time was £8?

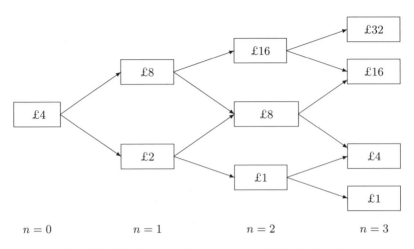

Fig. 13.1   The three-step binary price model for Problem 3.

4. Consider the case of the market model defined in Problem 3 above. Using your answer to Problem 3(i) and doing further necessary calculations, find the price at time $n = 0$ and the optimal stopping time τ_0^* for the American-style arithmetic-average Asian put option with strike $K = £16$ and maturity time $n = 3$.

5. Consider the price model from Problem 3 above. Assume that the 'actual' probability P for the price model in this example is such that each price movement at each step happens with probability $1/2$. Compute the Radon–Nikodym derivative Z of the risk-neutral measure Q found in Problem 3(i) with respect to the 'actual' measure P.

6. Let the three-step binary price model with the time step $\Delta t = 1/26$ (two weeks) have the structure shown in Fig. 13.2. Suppose the continuously compounded interest rate p.a. is such that $r = 26 \ln (5/4)$.

(i) Demonstrate that there are no arbitrage opportunities in this market model and find the risk-neutral probability for this model.

(ii) Consider the European call option with strike price equal to S_0 and maturity time $n = 3$.

(a) Find its value at time $t = 0$.

(b) Suppose the writer follows a perfect hedging strategy. What portfolio should he have at time $n = 2$ if the stock price at this time was £15?

(c) Find the value of the corresponding European put at $t = 0$ using the put–call parity.

(d) What are the values of the payoff of a European-type arithmetic-

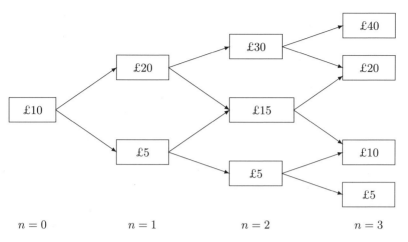

Fig. 13.2   The three-step binary price model for Problem 6.

average Asian call option with strike $K = S_0$ and maturity time $n = 3$ in the case of this model?

7. An American-style discrete *Russian option* has the payoff

$$f_n = \max_{0 \leq k \leq n} S_k, \quad n = 0, \ldots, N.$$

Consider the case of the market model defined in Problem 6 above and the maturity time $n = 3$. Using your answer to Problem 6(i) and doing further necessary calculations, find the price at time $n = 0$ and the optimal stopping time τ_0^* for this option.

8. Let the three-step binomial price model with the time step $\Delta t = 1/12$ have the structure shown in Fig. 13.3. Let the continuously compounded *variable* interest rate be such that $r_0 = 12 \ln 1.005$ during the first period, $r_1 = 12 \ln 1.004$ during the second period and $r_2 = 12 \ln 1.003$ during the third period.

(a) Are there any arbitrage opportunities in the market model? If so, find them.

(b) If there are no arbitrage opportunities in the market model, then

(1) find the value at time $t = 0$ of the European put option with strike price £80 and maturity time $1/4$.

(2) Find the writer's hedging portfolio at $t = 0$.

9. An American-style discrete *lookback option* has the payoff

$$f_n = \max_{0 \leq k \leq n} S_k - S_n, \quad n = 0, \ldots, N.$$

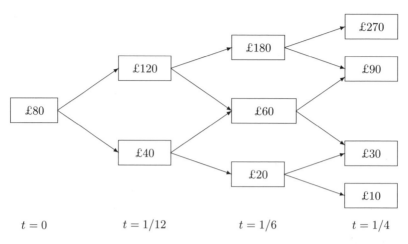

Fig. 13.3   The three-step binary price model for Problem 8.

Consider the case of the market model defined in Problem 8, above, and the maturity time $N = 3$. Using your answer to Problem 8, find the values of the hedging portfolio at time $n = 2$ depending on the information flow.

10. Using Problem 23 from Chapter 6, explain why the prices of a European put and the corresponding American put coincide when the interest rate is zero and the underlying stock does not pay a dividend.

11. Let $\xi_1, \ldots, \xi_N$ be a sequence of coin tosses and let $\mathcal{A}_n$, $n = 0, \ldots, N$, be the filtration generated by $\xi_1, \ldots, \xi_n$. For each of the following events, find the smallest $\mathcal{A}_n$ to which the event belongs (assume that N is sufficiently large):

(a) $A=$\{the first 7 tosses produce at least 3 tails\};

(b) $B=$\{the first occurrence of heads is preceded by no more than 9 tails\};

(c) $C=$\{there is at least one heads in the sequence $\xi_1, \xi_2, \ldots, \xi_N$\};

(d) $G=$\{there are no more than one heads and one tails among the first three tosses\}.

12. (*Probability revision*). Let ξ and η be some random variables with finite second moments.

(i) If ξ and η are independent and $a, b \in \mathbf{R}$ then prove that
$$Var(a\xi + \eta + b) = a^2 Var(\xi) + Var(\eta).$$

(ii) Covariance of two random variables ξ and ζ is defined as
$$Cov(\xi, \zeta) := E\left[(\xi - E\xi)(\zeta - E\zeta)\right] = E\left(\xi\zeta\right) - E\xi E\zeta.$$
Show that
$$Var(\xi + \eta) = Var(\xi) + Var(\eta) + 2Cov(\xi, \eta).$$

13. Toss a fair coin repeatedly, i.e. assume that the probability of getting a heads on each toss is $1/2$ as is the probability of getting a tails. Let $\zeta_i = 1$ if the ith toss results in a head and $\zeta_i = -1$ if the ith toss results in a tail. Consider the stochastic process η_n, $n = 0, \ldots, N$, defined by $\eta_0 = 0$ and $\eta_n = \sum_{i=1}^{n} \zeta_i$ and adapted to a filtration $\{\mathcal{A}_n\}$. This process is called a *symmetric random walk*; with each head it moves up by one, and with each tail it moves down by one. Show that η_n is a martingale with respect to the given probability measure.

14. Let ζ_i be independent identically distributed (i.i.d.) random variables with $P(\zeta_i = \pm 1) = 1/2$. Consider the stochastic process $\eta_n, n = 0, \ldots, N$, defined by $\eta_0 = 0$ and $\eta_n = \sum_{i=1}^{n} \zeta_i$ and its natural filtration $\{\mathcal{A}_n\}$. Show that

$$\theta_n = (-1)^n \cos(\pi \eta_n)$$

is a martingale with respect to the given probability measure and the filtration $\{\mathcal{A}_n\}$.

15. (*Discrete stochastic integral*). Let η_n, $n = 0, \ldots, N$, be a martingale with respect to a measure Q adapted to a filtration $\{\mathcal{A}_n\}$ and Δ_n, $n = 0, \ldots, N - 1$, be an $\{\mathcal{A}_n\}$-adapted process with $E|\Delta_n| < \infty$. We define the discrete-time stochastic integral $I_0, I_1, \ldots, I_N$ as[1] $I_n = \sum_{i=0}^{n-1} \Delta_i (\eta_{i+1} - \eta_i)$, $n = 0, \ldots, N$. Show that I_n is a martingale with respect to the measure Q.

16. (*Doob's optional stopping theorem*[2]). Suppose η_n, $n = 0, 1, 2, \ldots$, is a martingale with respect to a measure Q and adapted to a filtration $\{\mathcal{A}_n\}$. Let $\tau = \tau(\omega)$ be a bounded[3] stopping time with respect to algebras $\mathcal{A}_n$. Then $E\eta_\tau = E\eta_0$.

17. Consider a discrete random process

$$S_n = S_{n-1} - \mu S_{n-1} + \sigma S_{n-1} \xi_n,$$

where $\sigma > 0$ and $0 \leq \mu$ are constants so that $1 - \mu - \sigma \geq 0$, ξ_n are i.i.d. random variables with the law $P(\xi_n = \pm 1) = 1/2$ and $S_0 > 0$. Prove that S_n is a supermartingale. For which μ is it a martingale?

18. Prove (9.17).

19. Let Z be the Radon–Nikodym derivative of Q with respect to P as defined in Section 9.3. Show that $E_P \xi = E_Q (\xi/Z)$.

---

[1] Recall the usual convention that when i in the sum is less than $n - 1$, the quantity is equal to zero, i.e. here $I_0 = 0$.

[2] See its statement under weaker conditions in, e.g., Shiryaev (1996).

[3] i.e. there is an N such that $\tau(\omega) \leq N$ for all ω.

20. Assume that you can own only one type of stock S_n and one type of bond B_n which both follow the binomial model of price evolution $S_n = S_{n-1}(1 + \xi_n)$ and $B_n = B_{n-1}(1 + \zeta_n)$, where ξ_n and ζ_n are independent random variables taking two different values each with non-zero 'actual' probability. Is this market model complete?

21. Assume that the market has only one type of stock which is modelled by a trinomial tree

$$S_n = S_{n-1} + S_{n-1}\xi_n,$$

where ξ_n is a random variable which can take one of the three values $a = -0.5$, $b = 0.5$ and $c = 1$; and one type of bond which follows the deterministic dynamics: the value of one unit of bond at time $t_n = n\Delta t$ is $B_n = B_0 e^{rn\Delta t}$ with $r = \ln(1.05)/\Delta t$. Assume that the 'actual' probability P of ξ_n taking each of the three values is non-zero on every path and $S_0 = 10$.

(**a**) Confirm that this market model is arbitrage free and incomplete.

(**b**) Consider a European call on this underlying with strike $K = 5$ and maturity $T = \Delta t$. Can this option be replicated? Does it have a unique arbitrage price?

22. Assume that there is only one type of stock you can own and one type of bank account with continuously compounded interest rate $r = \ln(1.05)/\Delta t$ you can use. Suppose the stock price is modelled by a trinomial tree

$$S_n = S_{n-1} + S_{n-1}\xi_n,$$

where ξ_n is a random variable which can take one of the three values $a = -0.5$, $b = 0$ and $c = 0.5$. Assume that the 'actual' probability P of ξ_n, taking each of the three values, is non-zero on every path and $S_0 = 5$. Demonstrate that this market model is arbitrage free. Find the infimum and supremum of arbitrage prices for a European call on this underlying with the strike $K = 4$ and maturity $T = \Delta t = 1/52$.

23. Let $\varphi(x)$, $x \in \mathbf{R}$, be a concave function and $a_1, \ldots, a_n$ be some positive weights and $x_1, \ldots, x_n \in \mathbf{R}$. Prove the finite form of Jensen's inequality for *concave*[4] functions:

$$\varphi\left(\frac{\sum_{i=1}^n a_i x_i}{\sum_{i=1}^n a_i}\right) \geq \frac{\sum_{i=1}^n a_i \varphi(x_i)}{\sum_{i=1}^n a_i}.$$

---

[4]A function $U(s)$, $s \in \mathbf{R}$, is called a **concave** function if, for any $s_1, s_2 \in \mathbf{R}$ and $\lambda \in [0, 1]$, we have
$$U(\lambda s_1 + (1 - \lambda)s_2) \geq \lambda U(s_1) + (1 - \lambda)U(s_2).$$

An example of a concave function is: $U(s) = \ln s$ for $s > 0$ and $U(s) = -\infty$ for $s \leq 0$ which is often used as a risk aversion measure (a utility function) in, e.g., portfolio management (see, e.g., Levy, 2006).

24. Let $x_1, \ldots, x_n \in \mathbf{R}$. The arithmetic average is defined as $\frac{1}{n} \sum_{i=1}^{n} x_i$. Let $x_1, \ldots, x_n$ be some positive numbers. The geometric average is defined as $\sqrt[n]{x_1 x_2 \ldots, x_n}$. Using Jensen's inequality from Problem 23, prove that

$$\frac{1}{n} \sum_{i=1}^{n} x_i \geq \sqrt[n]{x_1 x_2 \ldots x_n}.$$

25. Consider ABC and XYZ shares for which the returns have variances σ_1^2 and σ_2^2, respectively, and correlation coefficient of the returns is ρ. You bought one ABC share and want to hedge your position by short selling x number of XYZ shares so that your portfolio has the minimum possible variance[5]. Find the value of x.

---

[5]Here, variance is viewed as a measure of risk which is typical for Markowitz's portfolio theory and in the capital asset pricing model (CAPM) developed by William F. Sharpe, Jack Treynor, John Lintner and Jan Mossin in the 1960s (resulting in a Nobel Prize for Markowitz and Sharpe in 1990).

Part III

Continuous-Time Stochastic Modelling and the Black–Scholes Formula

There're many uncertainties in this strange land,
We could get disoriented and confused. . .
Even shivers are running down our spine,
If we imagine what could possibly happen[1].

We start this part with some basic modelling consideration including the efficient market hypothesis. The main result of this final part of the course is the famous Black–Scholes formula for pricing European options. We will derive it in two ways. Firstly, we will obtain it as the limit of the discrete Black–Scholes formula from Chapter 10 via the application of the central limit theorem. Secondly, we will derive it by starting from a continuous-time price model (geometric Brownian motion). The latter requires us to enhance our knowledge of Stochastic Analysis, in particular by introducing the Wiener process, Ito integrals and stochastic differential equations.

---

[1]Translated from a song from Visotsky's audioplay; see Footnote 1 on page 5.

Chapter 14

Connection to 'Reality'

In the binary tree model (I) of Part II, the time period (step) Δt corresponds to some length of calendar time, e.g., a day, and at the end of it we only have two possible values. This is obviously far from reality. First of all, asset prices during a trading day usually take a lot of different values. Second, trading on the market is almost continuous. So, we have two objections for the model (I) but we can adjust it to account for that, as we will see in Chapter 16. Now let us briefly outline how we will proceed.

The obvious way of adapting the model is to make the time step Δt sufficiently small (e.g., corresponding to a minute or a second). This will address both objections. Indeed, asset prices can then take many values during the trading day and trades happen with desirable frequency. However, to match our model to the reality, we need to adjust it further. Intuitively, the smaller the step Δt the smaller the price change should be. Then we need to make a_n and b_n in the binary model (I) (see Chapter 7) dependent on the time step in a reasonable way from the applicable point of view, i.e. for sufficiently small Δt the model's outcome should reasonably correspond to the empirical evidence of asset price behaviour on the market[1]. If we achieve this, then we can use the binary model for practice and also gain an insight into continuous models.

Before we deal with the limit of model (I) as $\Delta t \to 0$, we will discuss actual price behaviour and also continue the study of Probability.

---

[1]This problem was first considered in Cox *et al.* (1979).

14.1 Efficient Market Hypothesis

The price of an asset is a measure of investors' confidence and belief in the quality of the asset. It depends on various economic data, political events, good or bad news about a company, etc., i.e. it depends on information which is related (directly or indirectly) to the asset and is available to market participants. The simplified assumption usually used in Finance is that the market responds instantaneously to the update of information, or in other words that 'The current asset price reflects all past information'. This assumption is called the **efficient market hypothesis**. It means that the complete history of the related information cannot improve prediction of the future asset price made on the basis of the current state of the market. In other words: '...*there is no edge to be gained from "reading the charts"*' (see p. 45 in Higham, 2004).

As any hypothesis, it is open for criticism[2]. It is also open to interpretation, e.g., what the full current information means or whether various market participants have the same information available to them (Kolb, 2003). Nevertheless, this hypothesis is used in almost all models for financial engineering purposes and in the risk-neutral world it takes the form of the so-called random walk hypothesis: the asset price evolution from time t to $t + \Delta t$ uses the asset price at time t only. This is so in the binary model (I) from Chapter 7 and we will also keep this assumption in our future considerations[3]. For further reading on the efficient market hypothesis, see Barucci (2003); Kolb (2003) and Shiryaev (1999).

14.2 Market Data and Model Assumptions

Exercise 14.1. Find market data on any asset (for instance, you can look on the Yahoo finance website at charts for, e.g., FTSE 100 index and British Airways shares or you can find similar charts on, e.g., commodities on the BBC Market data website. Analyse these data on different time scales, say, one month, three months, one year, etc.

---

[2]See, e.g., Barucci (2003) and Lo and MacKinlay (2002) and also the references therein.

[3]Such a property (future depends on the past only via the current state) of stochastic processes is known as the Markov property and the corresponding processes are called Markov processes. Here we avoid rigorous introduction of the Markov property and thus will not use these terms in the course. If you would like to familiarise yourself with the concept of Markov processes, you can find the corresponding material in, e.g., Shiryaev (1996); Shreve (2003, 2004) and Wentzell (1981).

If you did the above exercise, you could have observed similar jaggedness of data covering different time scales. If you looked at price behaviour on even a smaller scale, say within one trading day, you saw a similar type of price uncertainty. One can make a statistical analysis of this data in the following manner[4].

Consider the characteristic known as *daily return*:

$$R(t_i) := \frac{S(t_{i+1}) - S(t_i)}{S(t_i)},$$

where $S(t_i)$ and $S(t_{i+1})$ are the asset prices on consecutive trading days (on the charts you looked at, when doing Exercise 14.1, each point is usually the price at the last transaction made in each trading day; such prices are called *close-of-trading prices*). We note that the return $R(t_i)$ is a dimensionless quantity which is convenient for comparing various assets. We can compute *sample mean* $\tilde{\mu}_M$ and *sample variance* $\tilde{\sigma}_M^2$ as

$$\tilde{\mu}_M = \frac{1}{M} \sum_{i=1}^{M} R(t_i), \quad \tilde{\sigma}_M^2 = \frac{1}{M-1} \sum_{i=1}^{M} (R(t_i) - \tilde{\mu}_M)^2.$$

We recall from Statistics that sample mean and variance serve as estimators for the corresponding characteristics of random variables. Working with time series, cleverer estimates can be proposed and used. In addition, we implicitly assume stationarity of data but will avoid any detailed discussion here (see Footnote 4).

Now we normalise the daily returns as

$$\tilde{R}(t_i) := \frac{R(t_i) - \tilde{\mu}_M}{\tilde{\sigma}_M},$$

i.e. we subtract the estimated trend $\tilde{\mu}_M$ and normalise by the estimated standard deviation $\tilde{\sigma}_M$.

Exercise 14.2. (This one will take some time to complete and it, together with the next three exercises, can actually constitute a good project.) Take the data you found in Exercise 14.1, download them in an electronic spreadsheet and compute the corresponding normalised daily returns $\tilde{R}(t_i)$. Then produce the relative frequency histogram[5] and cumulative probability distribution function for the data obtained. Compare these graphs with the curves

---

[4]Note that statistical (data mining) methods are not considered in this course in any depth, we just briefly indicate here where they can potentially be used within our modelling framework. Analysis of data is an essential part of financial modelling and decision making. It is advisable that you study the corresponding methods. See also some warnings about the financial data on pp. 49–50 of Higham (2004).

[5]It is a graphical representation visualising the distribution of experimental data.

corresponding to the normal distribution $\mathcal{N}(0,1)$ with zero mean and unit variance[6]. In addition, using a random number generator, produce a sample from $\mathcal{N}(0,1)$. For this sample, plot the relative frequency histogram and cumulative probability distribution function and compare them with the ones for the market data. For further reading, see Chapter IV in Shiryaev (1999) and pp. 46–48 in Higham (2004).

Exercise 14.3. Repeat the previous exercise for weekly returns.

As it is widely accepted (see, e.g., Shiryaev (1999) and Higham (2004) and references therein), analysis like the one described in the above exercise suggests that normalised daily and weekly returns behave *approximately* in a similar manner to the sample from i.i.d. Gaussian random variables. Needless to say here that this is an approximation and how good or bad it is depends on the particular asset and situation during the considered time period.

Exercise 14.4. Can the above conclusion be accepted for the data you studied in the previous exercises?

Further, the daily returns are quite small and the *log-returns* $\ln \dfrac{S(t_{i+1})}{S(t_i)}$ can be approximated as

$$\ln \frac{S(t_{i+1})}{S(t_i)} = \ln \left(1 + \frac{S(t_{i+1}) - S(t_i)}{S(t_i)} \right) \approx \frac{S(t_{i+1}) - S(t_i)}{S(t_i)} = R(t_i).$$

Hence, as a rule, behaviour of log-returns is similar to the behaviour of simple returns.

The standard assumption for the average of log-returns over period Δt is

$$E \left[\ln \frac{S(t + \Delta t)}{S(t)} \right] = \mu \Delta t,$$

where μ is a constant parameter. Typically $\mu > 0$ and represents a general upward drift of the asset price (a trend). The parameter μ is usually called the *drift*. The expectation here is under 'actual' (in other words, market) measure.

Let us note that in our market model growth of money on the bank account, B_t, has the (continuous-time) form:

$$B_t = B_0 e^{rt}$$

---

[6]Here we suggest just to rely on the visualisation of the data. Of course, one can do such a comparison using statistical techniques, e.g., the Kolmogorov–Smirnov test.

and

$$\ln \frac{B_t}{B_0} = rt.$$

The parameter μ plays the same role in the stock price model as the interest rate r in the model for the 'time value of money' from Section 2.2.

The variance of log-returns over period Δt is often considered to be equal to $\sigma^2 \Delta t$:

$$Var\left(\ln \frac{S(t + \Delta t)}{S(t)}\right) = \sigma^2 \Delta t,$$

where $\sigma \geq 0$ is a constant parameter that determines the strength of the random fluctuation. In finance this is called **volatility**. These assumptions can be confirmed, to some extent, by statistical analysis of market data.

Exercise 14.5. Are these assumptions consistent with your data from previous exercises?

Summarising: we assume that:

$$\ln \frac{S(t + \Delta t)}{S(t)} \sim \mathcal{N}(\mu \Delta t, \sigma^2 \Delta t), \tag{14.1}$$

i.e. that log-returns of a stock over the time increment Δt are normally distributed with mean $\mu \Delta t$ and variance $\sigma^2 \Delta t$. We can rewrite (14.1) as

$$\ln S(t + \Delta t) = \ln S(t) + \mu \Delta t + \sigma \sqrt{\Delta t}\theta$$

or

$$S(t + \Delta t) = S(t) \exp\left(\mu \Delta t + \sigma \sqrt{\Delta t}\theta\right), \tag{14.2}$$

where $\theta \sim \mathcal{N}(0, 1)$, i.e. we can say that the stock price $S(t + \Delta t)$ is *lognormally distributed* assuming that the value of $S(t)$ is given[7]. We observe that $S(t + \Delta t)$ from (14.2) is positive if $S(t) > 0$.

A similar assumption was already mentioned, when we considered forward contracts in Section 3.1. It is widely used in pricing practice. In particular, an assumption of this kind was used by Samuelson (1965), when he introduced geometrical Brownian motion into financial modelling. Samuelson's model was later used by Merton (1973) and Black and Scholes (1973) to derive their Nobel Prize winning pricing formula. At the same time, this assumption is, of course, oversimplistic and open to criticism. Various modifications of this model exist but they are beyond the remit of this course (for further reading one can use, e.g., Musiela and Rutkowski (2005); Shreve (2004) and Kwok (2008)).

After we introduce a probabilistic model for experiments with infinitely many outcomes in the next chapter, we will link the binary model of price evolution (I) and the log-normal model (14.2) together.

---

[7]We slightly abuse rigour here, keeping the presentation on the intuitive level.

Chapter 15

Probabilistic Model for an Experiment with Infinitely Many Outcomes

15.1 Probabilistic Model[1]

The probabilistic models of Part II allowed us to describe experiments with a finite number of outcomes.

Example 15.1. Consider a random experiment of independently tossing a coin n times with probability p of heads (such an experiment is often called Bernoulli trials). The probabilistic model for this experiment is the triple $(\Omega, \mathcal{A}, P)$ with the sample space $\Omega = \{\omega : \omega = (a_1, \ldots, a_n), a_i = 0, 1\}$ (here '0' represents heads and '1' – tails), the algebra $\mathcal{A} = \{A : A \subseteq \Omega\}$ and elementary probabilities

$$p(\omega) = p^{\sum_{i=1}^n a_i} (1 - p)^{n - \sum_{i=1}^n a_i}.$$

The number of outcomes, i.e. the number of points in Ω, is finite and equal to 2^n.

How to extend this probabilistic model to the case of an *infinite* number of tosses of a coin? The natural extension of the sample space is

$$\Omega = \{\omega : \omega = (a_1, a_2, \ldots), \ a_i = 0, 1\},$$

i.e. the space of sequences $\omega = (a_1, a_2, \ldots)$, in which elements a_i are either 0 or 1. How many elements does this set have? As it is well known, any number $a \in [0, 1)$ has a unique binary expansion:

$$a = \frac{a_1}{2} + \frac{a_2}{2^2} + \cdots,$$

where a_i is either 0 or 1. Therefore, there is a one-to-one correspondence between the points of the Ω and the interval $[0, 1)$. Consequently, Ω is an

[1]Shiryaev (1996) was used in preparing this chapter.

uncountable set (uncountably infinite)[2]. We say that the cardinality[3] of the interval is continuum. So, we have the first element of the probabilistic model – the sample space Ω, which is rather complex.

Let us now consider how to introduce probabilities for this experiment with infinitely many outcomes. For simplicity, assume that the coin is fair, i.e. $p = 1/2$. Due to symmetry, all outcomes $\omega \in \Omega$ are equiprobable. However, as we have concluded, the sample space Ω is uncountable. Further, as usual, probability of the whole Ω is 1. Then the probability $p(\omega)$ of each $\omega \in \Omega$ is inevitably equal to 0! This assignment of probabilities $p(\omega) = 0$, $\omega \in \Omega$, is not useful, of course. What is the way out from this logical dead end?

In fact, we are usually interested not in probabilities of a particular outcome but in the probability that the result of the experiment is in a set A of outcomes (i.e. in an event). In elementary Probability theory that deals with finite probability spaces, we find the probability $P(A)$ of the event A (see Section 8.1) via elementary probabilities as

$$P(A) = \sum_{\omega \in A} p(\omega).$$

Clearly, when $p(\omega) = 0$, $\omega \in \Omega$, we cannot define $P(A)$ in this way.

We note that due to the observation made earlier in this section, the experiment of tossing a fair coin infinitely many times is equivalent to the experiment of choosing points from the interval $[0, 1)$ at random. Suppose the event A of our interest is $[0, 1/2)$. Then it is intuitively clear that the probability $P(A)$ of this A should be $1/2$. This common sense consideration suggests that in constructing a probabilistic model in the case of an uncountable sample space Ω, probabilities should be assigned to events rather than to individual outcomes.

Such a probabilistic model was constructed in an axiomatic way by A. N. Kolmogorov in 1933 (Kolmogorov, 1933). This was the fundamental discovery, which is the basis of the modern Probability theory, obviously including Stochastic Analysis and Financial Mathematics.

As in the case of a finite number of outcomes, we also need three elements in the probabilistic model for experiments with an infinite number of outcomes: a sample space Ω, which we have already discussed, a collection of events and a probability measure.

---

[2]That is, it is not possible to list its elements in a sequence. Set of natural numbers $\mathbb{N}$ is countable. If there is no one-to-one correspondence between $\mathbb{N}$ and an infinite set, then this set is uncountable.

[3]That is, the number of elements in a set.

Collection of events must be closed with respect to unions, intersections and compliments; and when we deal with an infinite number of outcomes, we have to take into account results of an infinite number of unions and intersections. To this end, consider the following definition.

Definition 15.1. *Let Ω be a non-empty set and $\mathcal{F}$ be a collection of subsets of Ω. We say that $\mathcal{F}$ is a σ-algebra provided that*
1. $\Omega \in \mathcal{F}$.
2. *If $A_n \in \mathcal{F}$, $n = 1, 2, \ldots$, then $\cup A_n \in \mathcal{F}$, $\cap A_n \in \mathcal{F}$.*
3. $A \in \mathcal{F} \Rightarrow \overline{A} \in \mathcal{F}$.

In other words, $\mathcal{F}$ is a σ-algebra if it is an algebra (see Definition 8.3 in Section 8.1) and satisfies this new (stronger than condition 2 in the definition of algebra) condition 2. Note that it is sufficient to require either $\cup A_n \in \mathcal{F}$ or $\cap A_n \in \mathcal{F}$ in Definition 15.1.

Definition 15.2. *The space Ω together with a σ-algebra $\mathcal{F}$ of its subsets is a **measurable space** $(\Omega, \mathcal{F})$.*

Example 15.2. (*The measurable space* $(\mathbf{R}, \mathcal{B}(\mathbf{R}))$). Let $\mathcal{A}$ be the system of subsets of the real line $\mathbf{R}$ consisting of finite unions of disjoint intervals of the form $(a, b]$ together with the empty set $\emptyset$:

$$A \in \mathcal{A} \text{ if } A = \bigcup_{i=1}^{n} (a_i, b_i], \quad n < \infty.$$

We can check (please do so) that $\mathcal{A}$ is an algebra. But it is not a σ-algebra. Indeed, if $A_n = (0, 1 - \frac{1}{n}] \in \mathcal{A}$ then $\bigcup_{n=1}^{\infty} A_n = (0, 1)$ which is not in $\mathcal{A}$.

Denote by $\mathcal{B}(\mathbf{R})$ the smallest σ-algebra $\sigma(\mathcal{A})$ containing $\mathcal{A}$. It is called the **Borel σ-algebra** of subsets of the real line and its sets are called **Borel sets**.

We have

$$(a, b) = \bigcup_{n=1}^{\infty} (a, b - \frac{1}{n}], \quad a < b;$$

$$[a, b] = \bigcap_{n=1}^{\infty} (a - \frac{1}{n}, b], \quad a < b; \quad \{a\} = \bigcap_{n=1}^{\infty} (a - \frac{1}{n}, a].$$

Then the Borel σ-algebra contains not only intervals of the form $(a, b]$ but also points (singletons) $\{a\}$ and all sets of the following forms

$$(a, b), \quad [a, b], \quad [a, b), \quad (-\infty, b), \quad (-\infty, b], \quad (a, \infty). \tag{15.1}$$

We also note that we could construct the Borel σ-algebra $\mathcal{B}(\mathbf{R})$ by starting with any of the six kinds of intervals from (15.1) instead of $(a, b]$.

Example 15.3. (*The measurable space* $([0,1], \mathcal{B}([0,1]))$). The Borel σ-algebra $\mathcal{B}([0,1])$ is the collection of Borel subsets of $[0,1]$:

$$\mathcal{B}([0,1]) = \{A \cap [0,1] : \ A \in \mathcal{B}(\mathbf{R})\}.$$

We now have two elements in our probabilistic model. We need the third – probability.

Definition 15.3. *Let* $(\Omega, \mathcal{F})$ *be a measurable space. A* ***probability measure*** $P = P(A)$ *is a function that assigns a number in* $[0,1]$ *to every set* $A \in \mathcal{F}$ *and satisfies the following axioms:*

1. $P(\Omega) = 1$;
2. *(**countable additivity**) whenever* $A_1, A_2, \ldots$ *is a sequence of pairwise disjoint sets in* $\mathcal{F}$, *then*

$$P\left(\bigcup_{n=1}^{\infty} A_n\right) = \sum_{n=1}^{\infty} P(A_n).$$

$P(A)$ *is called the* ***probability*** *of an event A.*

We can now formulate **Kolmogorov's** universally accepted **axiomatic system**.

Definition 15.4. *An ordered triple* $(\Omega, \mathcal{F}, P)$, *where*

1. Ω *is a set of points* ω,
2. $\mathcal{F}$ *is a* σ-*algebra of subsets of* Ω *and*
3. P *is a probability measure on* $\mathcal{F}$,
   is called a ***probabilistic model*** *or a* ***probability space***. *Here,* Ω *is the sample space or space of elementary events, the sets* A *in* $\mathcal{F}$ *are events and* $P(A)$ *is the probability of the event A.*

Example 15.4. Consider the measurable space $(\mathbf{R}, \mathcal{B}(\mathbf{R}))$. Let $P = P(A)$, $A \in \mathcal{B}(\mathbf{R})$, be a probability measure. For $A = (-\infty, x]$ we define:

$$F(x) := P(-\infty, x], \ \ x \in \mathbf{R}.$$

We can check (please do so) that the function $F(x)$ possesses the properties:

1) $F(x)$ is non-decreasing;
2) $F(-\infty) = 0$, $F(+\infty) = 1$, where $F(\pm\infty) = \lim_{x \to \pm\infty} F(x)$;
3) $F(x)$ is continuous on the right[4] and has a limit on the left at each $x \in \mathbf{R}$.

---

[4]A right-continuous function is a function which is continuous at all points when approached from the right.

Definition 15.5. *A function $F = F(x)$ satisfying the above conditions 1–3 is called a **distribution function** (on the real line **R**).*

Proposition 15.1. *Let $F = F(x)$ be a distribution function on the real line **R**. There exists a unique probability measure P on $(\mathbf{R}, \mathcal{B}(\mathbf{R}))$ such that $P(a, b] = F(b) - F(a)$ for all $-\infty \le a, b < \infty$.*

We see that there is a one-to-one correspondence between probability measures P on $(\mathbf{R}, \mathcal{B}(\mathbf{R}))$ and distribution functions F on the real line **R**. The measure P constructed from the function F is usually called the **Lebesgue–Stieltjes** probability measure corresponding to the distribution function F.

Example 15.5. (*Uniform (Lebesgue) measure on* $[0, 1]$). Let us return to the problem with which we started this section and construct a mathematical model for choosing a number at random from the unit interval $[0, 1]$ so that the probability is distributed uniformly over the interval. We define the probability of an interval $(a, b] \subseteq [0, 1]$ as

$$P(a, b] = b - a, \quad 0 \le a \le b \le 1, \tag{15.2}$$

i.e. the probability that the number chosen between a and b is $b - a$. It is not difficult to understand that $P(a, b]$ satisfies Definition 15.3. This particular probability measure on $[0, 1]$ is called the **Lebesgue measure** on $[0, 1]$. The Lebesgue measure of a subset of **R** is its 'length'. In this case the distribution function is:

$$F(x) = \begin{cases} 0, & x < 0, \\ x, & 0 \le x \le 1, \\ 1, & x > 1. \end{cases}$$

The Lebesgue measure of any of the intervals (a, b), $[a, b]$ or $[a, b)$ is $b - a$ as well. We have completed the construction of the required probabilistic model – $([0, 1], \mathcal{B}([0, 1]), P)$ with P from (15.2).

From now on, we assume that we have a sufficiently rich probability space $(\Omega, \mathcal{F}, P)$ to describe our experiments.

15.2   Random Variables: Revisited

Here we revisit the notion of a random variable. We start with the definition of a random variable which is similar to Definition 8.6 of random variables in the discrete case.

Definition 15.6. *A real-valued **random variable** $\xi = \xi(\omega)$ is a real-valued function on Ω that is $\mathcal{F}$-measurable that means we require*

$$\{\omega : \xi(\omega) \in B\} \in \mathcal{F} \tag{15.3}$$

for every $B \in \mathcal{B}(\mathbf{R})$.

Remark 15.1. As before, we can view a random variable as a qualitative characteristic of an experiment, of which value depends on an outcome $\omega \in \Omega$ (or in the common language, on 'chance'). Here the requirement (15.3) is due to the following reason. Recall that a probability measure P is defined on $(\Omega, \mathcal{F})$. Then we need (15.3) in order to be able to assign the probability to the event $\{\xi(\omega) \in B\}$, i.e. that the value of the random variable belongs to the Borel set B; otherwise, random variables would not be of much use.

We now revise some of characteristics of random variables which we considered in the discrete case in Section 8.2.

Definition 15.7. *Let a random variable ξ be defined on $(\Omega, \mathcal{F})$ and consider a probability space $(\Omega, \mathcal{F}, P)$. A probability measure P_ξ on $(\mathbf{R}, \mathcal{B}(\mathbf{R}))$ defined as*

$$P_\xi(B) = P\{\omega : \xi(\omega) \in B\}, \quad B \in \mathcal{B}(\mathbf{R}),$$

*is called the **probability distribution** of ξ on $(\mathbf{R}, \mathcal{B}(\mathbf{R}))$.*
   The function

$$F_\xi(x) = P\{\omega : \xi(\omega) \leq x\}, \quad x \in \mathbf{R},$$

*is called **the distribution function** of ξ.*

In the next definition we introduce a classification of random variables.

Definition 15.8. *A random variable ξ is called*

- ***discrete** if it takes values in a countable set $X = \{x_1, x_2, \ldots\}$;*
- ***continuous** if its distribution F_ξ is continuous for $x \in \mathbf{R}$;*
- ***absolutely continuous** if there is a non-negative function $\rho = \rho_\xi(x)$, called its **density**, such that*

$$F_\xi(x) = \int_{-\infty}^{x} \rho_\xi(z)dz, \quad x \in \mathbf{R}.$$

Example 15.6. Gaussian random variable is an absolutely continuous random variable with the density

$$\rho(x) = \frac{1}{\sqrt{2\pi\sigma^2}} \exp\left(-\frac{(x-\mu)^2}{2\sigma^2}\right)$$

and the distribution function

$$\Phi(x; \mu, \sigma) = \frac{1}{\sqrt{2\pi\sigma^2}} \int_{-\infty}^{x} \exp\left(-\frac{(z-\mu)^2}{2\sigma^2}\right) dz.$$

Exercise 15.1. Find an example of a random variable which is continuous but not absolutely continuous[5].

To rigorously introduce expectation (which will incorporate all cases – discrete, continuous, absolute continuous), an introduction of the Lebesgue integral is needed (see, e.g., Shiryaev, 1996). This will be skipped in this course. We know already how to compute expectation in the discrete case (see Definition 8.8 in Section 8.2). In the *absolute continuous case* we define the **expectation** of ξ via the Riemann integral as

$$E\xi := \int_{-\infty}^{\infty} x\rho_\xi(x)dx$$

or more generally

$$Eg(\xi) := \int_{-\infty}^{\infty} g(x)\rho_\xi(x)dx,$$

where ρ_ξ is the density of ξ and $g(x)$ is a 'good' function.

Example 15.7. Let ξ be a Gaussian random variable with the density from Example 15.6. Then $E\xi = \mu$, $Var\xi = \sigma^2$.

Remark 15.2. Expectation and variance of continuous random variables have the same properties as in the discrete case (see Section 8.2).

Now we extend the definition of independence of random variables (see also Definition 8.10 in Section 8.3).

Definition 15.9. *We call random variables $\eta_1, \eta_2, \ldots, \eta_n$ **independent** if any of the following holds:*

1. For any n and any $x_1, x_2, \ldots, x_n$: the joint distribution function

$$F_{\eta_1,\ldots,\eta_n}(x_1, \ldots, x_n) = P(\eta_1 \leq x_1, \ldots, \eta_n \leq x_n)$$
$$= \prod_{k=1}^{n} P(\eta_k \leq x_k) = \prod_{k=1}^{n} F_{\eta_k}(x_k).$$

---

[5]You can find a solution to this exercise in e.g., Klebaner (2005) and Shiryaev (1996).

2. *In the case of absolutely continuous random variables defining the joint density*

$$\rho_{\eta_1,\ldots,\eta_n}(x_1,\ldots,x_n) := \frac{\partial^n F_{\eta_1,\ldots,\eta_n}(x_1,\ldots,x_n)}{\partial x_1,\ldots \partial x_n},$$

we get

$$\rho_{\eta_1,\ldots,\eta_n}(x_1,\ldots,x_n) = \prod_{k=1}^{n} \rho_{\eta_k}(x_k).$$

For the case of absolutely continuous random variables, the equivalence of 1) and 2) can be proved.

Example 15.8. Consider n independent Gaussian random variables with means m_i and variances σ_i^2. Their joint density has the form

$$\rho(x_1,\ldots,x_n) = \prod_{i=1}^{n} \frac{1}{\sqrt{2\pi\sigma_i^2}} \exp\left(-\frac{(x_i - m_i)^2}{2\sigma_i^2}\right)$$

$$= \frac{1}{(2\pi)^{\frac{n}{2}} \prod_{i=1}^{n} \sigma_i} \exp\left\{-\sum_{i=1}^{n} \frac{(x_i - m_i)^2}{2\sigma_i^2}\right\}.$$

Chapter 16

Limit of the Discrete-Price Model and Price of a European Option in the Continuous-Time Case

Let us recall that in Chapter 14 we decided to make a more-or-less realistic assumption that prices are log-normally distributed:

$$\ln \frac{S(t+s)}{S(t)} \sim \mathcal{N}(\mu s, \sigma^2 s), \quad s > 0, \tag{16.1}$$

with constant drift μ and volatility σ^2,

$$E\left[\ln \frac{S(t)}{S(0)}\right] = \mu t, \quad Var\left[\ln \frac{S(t)}{S(0)}\right] = \sigma^2 t. \tag{16.2}$$

Now let us return to our discrete binary model (I) from Chapter 7 with the price process $S_n = S_{n-1}(1 + \xi_n)$ and consider a sensible choice of parameters for it, sensible in terms of dependence of a_n and b_n on Δt as we discussed at the beginning of Chapter 14. Because we assumed above that μ and σ^2 are constant in (16.1), we will restrict ourselves to the binomial model, i.e. to the model (I) with $a_n \equiv a$, $b_n \equiv b$ and $p_n \equiv p$ for all n. In this model over each period Δt the stock price grows $S_n = S_{n-1}(1+b)$ with probability p and decreases $S_n = S_{n-1}(1 + a)$ with probability $1 - p$. We assume that the random variables ξ_n, $n = 1, \dots, N$, are independent under the 'actual' (market) measure P.

In other words, we would now like to choose a, b, p so that as $\Delta t \to 0$ our model (I) appropriately approximates what we consider to be the realistic assumption (16.1) on price behaviour.

Example 16.1. Show that if S_n follows the model (I) with $a_n \equiv a$, $b_n \equiv b$ and all the 'actual' probabilities of moving 'up' are $p_n \equiv p$ then

$$\bar{\mu}_n := E_P\left(\ln \frac{S_n}{S_0}\right) = \left[p \ln\left(\frac{1+b}{1+a}\right) + \ln(1+a)\right] n$$

and

$$\bar{\sigma}_n^2 := Var_P\left(\ln \frac{S_n}{S_0}\right) = p(1-p)\left[\ln\left(\frac{1+b}{1+a}\right)\right]^2 n.$$

167

Answer. Introduce

$$\zeta_n := \ln \frac{S_n}{S_{n-1}} = \ln(1 + \xi_n), \quad n = 1, \ldots, N.$$

We calculate:

$$E_P \zeta_n = E_P \ln(1 + \xi_n) = p \ln(1 + b) + (1 - p) \ln(1 + a)$$
$$= p \ln \left(\frac{1 + b}{1 + a} \right) + \ln(1 + a),$$

and (please check!)

$$Var_P \zeta_n = Var_P \ln(1 + \xi_n) = p(1 - p) \left[\ln \left(\frac{1 + b}{1 + a} \right) \right]^2.$$

We have

$$\ln \frac{S_n}{S_0} = \ln \left(\frac{S_1}{S_0} \frac{S_2}{S_1} \cdots \frac{S_{n-1}}{S_{n-2}} \frac{S_n}{S_{n-1}} \right) = \sum_{i=1}^{n} \ln \frac{S_i}{S_{i-1}} = \sum_{i=1}^{n} \zeta_i. \qquad (16.3)$$

Then

$$E_P \left(\ln \frac{S_n}{S_0} \right) = E_P \sum_{i=1}^{n} \zeta_i \stackrel{\text{property of expectations}}{=} \sum_{i=1}^{n} E_P \zeta_i$$
$$= \left[p \ln \left(\frac{1 + b}{1 + a} \right) + \ln(1 + a) \right] n,$$

as required.

Since the random variables ξ_n, $n = 1, \ldots, N$, are mutually independent, the random variables ζ_n, $n = 1, \ldots, N$, are mutually independent as well. Then

$$Var_P \left(\ln \frac{S_n}{S_0} \right) = Var_P \left(\sum_{i=1}^{n} \zeta_i \right) \stackrel{\text{independence}}{=} \sum_{i=1}^{n} Var_P \zeta_i$$
$$= p(1 - p) \left[\ln \left(\frac{1 + b}{1 + a} \right) \right]^2 n,$$

as required.

Let us go back to our discussion in Chapter 14. We were considering the division of a given fixed period T (e.g., a month or a year) into N shorter time periods Δt (a minute or even less), i.e. $T = N \Delta t$ and the shorter Δt the larger N. In searching for a realistic result, we must (as we have already discussed in Chapter 14) make the appropriate adjustments in a, b and p. In doing so, we would want, at least, that the mean $\bar{\mu}_N$ and variance $\bar{\sigma}_N^2$

coincide with that of the 'actual' (under our reasonable though simplistic model assumptions) stock price, expressed in (16.2).

Because the distribution in (16.1) is symmetric around μ, it is natural to take $p = 1/2$ (the price moves up with the same probability as it moves down).

Example 16.2. Show that if we choose a and b so that

$$a = \exp(\mu\Delta t - \sigma\sqrt{\Delta t}) - 1, \qquad (16.4)$$
$$b = \exp(\mu\Delta t + \sigma\sqrt{\Delta t}) - 1,$$

and $p = 1/2$ then

$$\bar{\mu}_N = \mu T \quad \text{and} \quad \bar{\sigma}_N^2 = \sigma^2 T. \qquad (16.5)$$

Answer. We have from Example 16.1:

$$\bar{\mu}_N = \left[p \ln\left(\frac{1+b}{1+a}\right) + \ln(1+a) \right] N.$$

Due to (16.4), we get

$$\ln(1+a) = \mu\Delta t - \sigma\sqrt{\Delta t}, \quad \ln(1+b) = \mu\Delta t + \sigma\sqrt{\Delta t}.$$

Then

$$\bar{\mu}_N = \left[\frac{1}{2}\left(\mu\Delta t + \sigma\sqrt{\Delta t}\right) - \frac{1}{2}\left(\mu\Delta t - \sigma\sqrt{\Delta t}\right) + \mu\Delta t - \sigma\sqrt{\Delta t} \right] N$$
$$= \mu\Delta t N = \mu T.$$

Analogously, we obtain

$$\bar{\sigma}_N^2 = p(1-p)\left[\ln\left(\frac{1+b}{1+a}\right) \right]^2 N$$
$$= \frac{1}{4}\left[\left(\mu\Delta t + \sigma\sqrt{\Delta t}\right) - \left(\mu\Delta t - \sigma\sqrt{\Delta t}\right) \right]^2 N$$
$$= \sigma^2 \Delta t N = \sigma^2 T.$$

Conclusion. Under the choice (16.4) of a and b and $p = 1/2$, we matched the drift and variance of our discrete model with those of the assumed 'actual' continuous price model.

Now let us discover what happens if we tend the time step Δt to zero in our discrete model (I) with the choice of parameters (16.4). Based on the numerical analysis experience, one would say that in the limit we shall obtain results corresponding to the continuous model. In order to study such limits, we need to revise the central limit theorem.

16.1 Central Limit Theorem and its Application

Central limit theorem (CLT)

Let $\vartheta_1, \vartheta_2, \ldots$ be a sequence of i.i.d. random variables under the probability measure P with a finite mean m and a finite non-zero variance ν^2 and let

$$\theta_n := \vartheta_1 + \cdots + \vartheta_n.$$

Then

$$\hat{\theta}_n := \frac{\theta_n - nm}{\sqrt{n\nu^2}}$$

converges in distribution to an $\mathcal{N}(0,1)$ random variable as $n \to \infty$ (which we write in short as $\hat{\theta}_n \implies \theta \sim \mathcal{N}(0,1)$), i.e. for any $-\infty \leq a < b \leq \infty$

$$P(a < \hat{\theta}_n < b) \to \Phi(b) - \Phi(a), \tag{16.6}$$

where $\Phi(x)$ is the normal distribution function given by

$$\Phi(x) = \frac{1}{\sqrt{2\pi}} \int_{-\infty}^{x} e^{-\frac{z^2}{2}} \, dz.$$

In a concise way, we can say that the distribution of $\hat{\theta}_n$ tends to the normal distribution with zero mean and unit variance when n goes to infinity. You can find a proof of CLT in, e.g., Shiryaev (1996).

Example 16.3. Let

$$\zeta_n := \ln \frac{S_n}{S_{n-1}}.$$

Due to the assumptions made before Example 16.1, the random variables ξ_n, $n = 1, \ldots, N$, in the model (I) are independent under the measure P, consequently $\dfrac{S_n}{S_{n-1}}$ are independent and, hence, ζ_n are independent under the measure P. Further (see Example 16.2):

$$E_P \zeta_n = \mu \Delta t, \quad Var_P \zeta_n = \sigma^2 \Delta t. \tag{16.7}$$

Introduce

$$\vartheta_n := \frac{\zeta_n - \mu \Delta t}{\sigma \sqrt{\Delta t}}$$

and note that

$$\zeta_n = \mu \Delta t + \sigma \sqrt{\Delta t} \vartheta_n.$$

It is clear that ϑ_n are independent and $E_P \vartheta_n = 0$ and $Var_P \vartheta_n = E_P \vartheta_n^2 = 1$ (please check).

Further, introduce

$$\theta_n := \sum_{i=1}^{n} \vartheta_i \quad \text{and} \quad \hat{\theta}_n := \frac{\theta_n}{\sqrt{n}}.$$

Due to CLT, we obtain that $\hat{\theta}_n$ converges in distribution to a Gaussian random variable with zero mean and unit variance:

$$\hat{\theta}_n \Rightarrow \tilde{\theta} \sim \mathcal{N}(0,1) \quad \text{as} \quad n \to \infty.$$

We have (recall that $T = N\Delta t$):

$$\ln \frac{S_N}{S_0} \overset{(16.3)}{=} \sum_{i=1}^{N} \zeta_i = \mu \Delta t N + \sigma \sqrt{\Delta t} \sum_{i=1}^{N} \vartheta_i \tag{16.8}$$

$$= \mu T + \frac{\sigma \sqrt{\Delta t} \sqrt{N}}{\sqrt{N}} \sum_{i=1}^{N} \vartheta_i$$

$$= \mu T + \sigma \sqrt{T} \hat{\theta}_N \Rightarrow \mu T + \sigma \sqrt{T} \tilde{\theta} \quad \text{as} \quad N \to \infty.$$

Hence,

$$\ln \frac{S_N}{S_0} \Rightarrow \theta \sim \mathcal{N}(\mu T, \sigma^2 T) \quad \text{as} \quad N \to \infty, \tag{16.9}$$

i.e. the log-return converges in distribution to a Gaussian random variable with mean μT and variance $\sigma^2 T$. Equivalently, we can write that

$$S_N \Rightarrow S_0 \exp(\theta) \tag{16.10}$$

with $\theta \sim \mathcal{N}(\mu T, \sigma^2 T)$.

Conclusion. As Δt gets smaller and N gets larger, the distribution of S_N (defined by (I) with the parameters from (16.4)) becomes log-normal as it is assumed for 'actual' price behaviour $S(T)$ in (16.1). In other words, we can say that for a small time step Δt and with parameters from (16.4), the binomial price model can be viewed as an approximation of the model in which the price is log-normally distributed with given mean and variance.

The next question is what happens to the price of a European option in such a limit.

16.2 Continuous Black–Scholes Formula

The result stated in (16.6) of CLT is equivalent to

$$E_P g(\hat{\theta}_n) \to E_P g(\tilde{\theta}) \quad \text{as} \quad n \to \infty, \tag{16.11}$$

where $\tilde{\theta} \sim \mathcal{N}(0,1)$ for any, e.g., continuous and bounded function g. This, of course, can be rigorously proved (see, e.g., Theorem 1, Section 1, Chapter 3 of Shiryaev (1996)).

We learned in Part II that in computing the fair price of a derivative we should use the risk-neutral probability. Under the assumptions made earlier in this chapter, the value of the risk-neutral probability for a stock price to go 'up' (see (II) from Section 9.2) is equal to (remember that we are considering the binomial tree model (I) with the coefficients defined in (16.4)):

$$q = \frac{e^{r\Delta t} - (1+a)}{b-a} = \frac{e^{r\Delta t} - e^{\mu\Delta t - \sigma\sqrt{\Delta t}}}{e^{\mu\Delta t}(e^{\sigma\sqrt{\Delta t}} - e^{-\sigma\sqrt{\Delta t}})} = \frac{e^{(r-\mu)\Delta t} - e^{-\sigma\sqrt{\Delta t}}}{e^{\sigma\sqrt{\Delta t}} - e^{-\sigma\sqrt{\Delta t}}}.$$

Using the Taylor expansion (we assume that the time step Δt is sufficiently small), we obtain

$$q = \frac{(r-\mu)\Delta t + \sigma\sqrt{\Delta t} - \frac{\sigma^2}{2}\Delta t + O((\Delta t)^{\frac{3}{2}})}{2\sigma\sqrt{\Delta t} + O((\Delta t)^{\frac{3}{2}})} \tag{16.12}$$

$$= \frac{1}{2}\frac{\sigma + (r - \mu - \frac{\sigma^2}{2})\sqrt{\Delta t} + O(\Delta t)}{\sigma + O(\Delta t)} = \frac{1}{2}\left[1 - \sqrt{\Delta t}\frac{\mu + \frac{\sigma^2}{2} - r}{\sigma}\right] + O(\Delta t)$$

with

$$|O((\Delta t)^\gamma)| \le K\Delta t^\gamma, \ \gamma > 0,$$

where $K > 0$ is a constant independent of Δt.

Analogously, the value of the risk-neutral probability for a stock price to go 'down' is

$$1 - q = \frac{1}{2}\left[\sqrt{\Delta t}\frac{\mu + \frac{\sigma^2}{2} - r}{\sigma} + 1\right] + O(\Delta t).$$

Under the risk-neutral probability Q, the random variables $\zeta_n = \ln\frac{S_n}{S_{n-1}}$ remain mutually independent (recall that we are dealing with the binomial tree), but now (please check and cf. (16.7)):

$$E_Q\zeta_n = (r - \frac{\sigma^2}{2})\Delta t + O((\Delta t)^{3/2}), \tag{16.13}$$

$$Var_Q\zeta_n = \sigma^2\Delta t + O((\Delta t)^{3/2}).$$

Introduce again

$$\vartheta_n := \frac{\zeta_n - E_Q\zeta_n}{\sqrt{Var_Q\zeta_n}},$$

which are i.i.d. under the probability Q with $E_Q \vartheta_n = 0$ and $Var_Q \vartheta_n = 1$. We have

$$\ln \frac{S_N}{S_0} = \sum_{i=1}^{N} \zeta_i = \sum_{i=1}^{N} \left(E_Q \zeta_i + \sqrt{Var_Q \zeta_i} \vartheta_i \right) = \sum_{i=1}^{N} E_Q \zeta_i + \sum_{i=1}^{N} \sqrt{Var_Q \zeta_i} \vartheta_i$$

$$\stackrel{(16.13)}{=} (r - \frac{\sigma^2}{2}) \Delta t N + O(N(\Delta t)^{3/2})$$

$$+ (\sigma \sqrt{\Delta t} + O(\Delta t)) \times \sqrt{N} \frac{\sum_{i=1}^{N} \vartheta_i}{\sqrt{N}}$$

$$= (r - \frac{\sigma^2}{2})T + O(\sqrt{\Delta t}) + (\sigma \sqrt{T} + O(\sqrt{\Delta t})) \tilde{\theta}_N,$$

where

$$\tilde{\theta}_N := \frac{\sum_{i=1}^{N} \vartheta_i}{\sqrt{N}}.$$

As $\Delta t \to 0$ (equivalently, $N \to \infty$, but remember that the time $T = N \Delta t$ is fixed), $O(\sqrt{\Delta t}) \to 0$ and, due to CLT, $\tilde{\theta}_N \Rightarrow \tilde{\theta} \sim \mathcal{N}(0,1)$. Therefore

$$\ln \frac{S_N}{S_0} \Rightarrow (r - \frac{\sigma^2}{2})T + \sigma \sqrt{T} \tilde{\theta} \quad \text{as} \quad N \to \infty,$$

where $\tilde{\theta} \sim \mathcal{N}(0,1)$. Hence

$$\ln \frac{S_N}{S_0} \Rightarrow \theta \sim \mathcal{N}((r - \frac{\sigma^2}{2})T, \sigma^2 T) \quad \text{as} \quad N \to \infty, \tag{16.14}$$

i.e. again the distribution of the log-return converges to the Gaussian distribution with variance $\sigma^2 T$ but with the mean $(r - \sigma^2/2)T$. It follows from (16.14) that

$$S_N \Rightarrow S_0 \exp(\theta)$$

with $\theta \sim \mathcal{N}((r - \frac{\sigma^2}{2})T, \sigma^2 T)$.

It is interesting and important that in the case of risk-neutral probability the limit does not depend on the drift μ – the 'real' trend of the price (see also Remark 16.2 below).

We had (see Theorem 10.1) the discrete Black–Scholes formula:

$$x^{\Delta t} = e^{-rT} E_Q[f(S_N)].$$

Due to (16.11) and (16.14), and assuming that f is 'good', we obtain under the special choice of the parameters a and b from (16.4):

$$x^{\Delta t} \to x = e^{-rT} E_Q[f(S_0 e^{\theta})] \quad \text{as} \quad \Delta t \to 0, \tag{16.15}$$

where $\theta \sim \mathcal{N}((r - \frac{\sigma^2}{2})T, \sigma^2 T)$. The formula (16.15) gives the price of a European-type option with payoff $f(s)$ in the continuous case. It is the famous *continuous* **Black–Scholes formula** for a European-type option.

Remark 16.1. We derived the Black–Scholes formula (16.15) as a special limiting case of our binary model (I). This derivation[1] required quite elementary mathematics in comparison with the original derivation via Ito Calculus and stochastic differential equations. Black and Scholes, and Merton began directly with continuous trading and the assumption of a log-normal distribution for stock prices. It is important to note that the economic arguments we used to link the option value and the stock price are the same as those assumed by Black and Scholes (1973) and Merton (1973) in their Nobel Prize winning works.

Example 16.4. Find the price of a European call in the continuous case.

Answer. Due to the Black–Scholes formula (16.15), we have

$$C_0 = e^{-rT} E_Q (S_0 e^\theta - K)_+$$

$$= e^{-rT} E_Q \left(S_0 \exp \left\{ (r - \frac{\sigma^2}{2})T + \sigma \sqrt{T} \tilde{\theta} \right\} - K \right)_+$$

$$= E_Q \left(S_0 \exp \left\{ -\frac{\sigma^2}{2}T + \sigma \sqrt{T} \tilde{\theta} \right\} - e^{-rT} K \right)_+ ,$$

where $\tilde{\theta} \sim \mathcal{N}(0, 1)$. After some relatively long calculations, the expectation can be rewritten in the form

$$C_0 = S_0 \Phi \left(\frac{\ln \frac{S_0}{K} + (r + \frac{\sigma^2}{2})T}{\sigma \sqrt{T}} \right) - K e^{-rT} \Phi \left(\frac{\ln \frac{S_0}{K} + (r - \frac{\sigma^2}{2})T}{\sigma \sqrt{T}} \right), \quad (16.16)$$

where $\Phi(x)$ is the normal distribution function.

Remark 16.2. The equation (16.15) (see also (16.16)) implies that the option price depends on the interest rate r (usually known *a priori*) and on the volatility σ of the price process – the only parameter we then need to estimate/calibrate to price options. This is important in practice: the volatility σ may be determined more easily and accurately than the trend μ.

To summarise, we obtained the limit of the discrete pricing formula, the Black–Scholes formula, we obtained the limiting distribution of $S(T)$ (and

---

[1]Apparently, it was first considered in Cox *et al.* (1979).

could do for $S(t)$ at any fixed t), which is log-normal and coincides with our 'realistic' assumptions on price behaviour. However, we have not considered what the limiting price *process* $S(t)$ is when $\Delta t \to 0$. To answer this question, we will need to first consider Brownian motion (Wiener process) which we introduce in the next chapter.

16.3   Estimation vs Calibration and Implied Volatility

As we observed and discussed in the previous section, to price a derivative according to the Black–Scholes formula (16.15) we need to estimate the single unobserved parameter – the volatility σ. Following our discussion in Section 14.2, it is natural to use historical price data for the underlying stock of the derivative for this task and use a favourite estimator of standard deviation for the time series to get the required value $\hat{\sigma}$ of volatility, which is called *historical volatility*.

However, in practice, historical volatility is, as a rule, not used for pricing derivatives. Instead, the so-called *implied volatility* is widely adopted in the financial industry for this purpose. Implied volatility is evaluated in the following way. Suppose there is a European-type derivative traded on the market, of which the underlier is the stock whose volatility we need to find. Then we can observe the current option price x and invert the Black–Scholes formula (16.15) to find the implied volatility σ_{imp} (recall that all the other parameters in (16.15) are known: the interest rate r and the stock's spot price are observable and the payoff of the derivative is given), which can be viewed as the value of volatility that was implicitly used by the market for pricing this benchmark option[2]. This procedure of finding volatility is called *calibration* to distinguish it from the estimation of σ based on historical data. Usually, if we need to price an OTC derivative with maturity T (say, one year), then, as a 'benchmark' used in calibration, one takes an observable derivative on the same underlier and with the same maturity.

The reasoning for using implied volatility rather than historical volatility is as follows. The real volatility is obviously changing with time. Historical volatility reflects market behaviour and expectations in the past whilst in pricing a derivative we need the best information possible about future volatility (e.g., over the next year as in the example in the previous para-

---

[2]We have met here with another 'implied' quantity. Roughly speaking, the 'implied' logic we have experienced in this course is what makes Finance different to Physics, Chemistry, Biology and, in fact, other parts of Economics in their approaches to and views on data analysis and interpretation.

graph). Current prices on the corresponding tradable options reflect traders' anticipation of what will happen with this particular underlying stock in the future and thus incorporate current and past information on which such a prognosis can be made. So, in other words, implied volatility is a better estimate for future volatility than its historical counterpart. From another angle, we want to price our OTC derivative in a consistent way with how the market prices other derivatives on the same underlier, which obviously requires the use of implied volatility.

This all looks nice and logical but in reality there is a problem with implied volatilities due to a deficiency in the Black–Scholes world. On a market we can usually observe a number of calls and puts written on the same stock but having different maturities and different strikes. According to the Black–Scholes formula, we should be able to price all of them with the same volatility σ or, in other words, all these options should have the same implied volatility σ_{imp}. However, the reality is different. Empirical data suggest that σ_{imp} depends on values of strikes K: usually options, which are far OTM or deep ITM, are traded at higher implied volatilities than ATM options (Björk, 2004). Tomas Björk says (see p. 104 in Björk, 2004): 'The graph of the observed implied volatility function thus often looks like the smile of the Cheshire cat, and for this reason the implied volatility curve is termed the *volatility smile*'. Implied volatilities also depend on the maturity time: σ_{imp} is usually larger for larger T, which reflects the intuitively obvious fact that the uncertainty of the market is higher over a larger time horizon. The function $\sigma_{imp}(K, T)$ is called the *volatility surface*. For further information on implied volatilities and calibration, see, e.g., Bingham and Kiesel (2004); Musiela and Rutkowski (2005) and Fengler (2009).

To fix this deficiency in the Black–Scholes formula, a number of more complicated models for stock were introduced, which include local volatility and stochastic volatility models and models with jumps (see, e.g., Bingham and Kiesel, 2004; Musiela and Rutkowski, 2005; Lamberton and Lapeyre, 2007; Shreve, 2004). To consider such models, we need more sophisticated mathematical tools, in particular stochastic differential equations, which will be our next topic. At the same time, we should note that whilst a simple model does not adequately represent the reality, a 'perfect' model might be impossible to calibrate due to lack of data or it might be impractical to calibrate due to its high computational costs. A model in Financial Engineering is of practical value only if it has a feasible calibration procedure associated with it. Emanuel Derman writes:

'A model is only a model; you want it to capture the essence of the phenomenon, not the thing itself. It is far too easy, in the name of realism, to add complexity to the simple evolution of stock prices assumed by Black and Scholes, but complexity without calibration is pointless.' (Derman, 2004, p. 231)

Chapter 17

Brownian Motion (Wiener Process)

In 1827 Scottish botanist Robert Brown reported that he observed an irregular motion of pollen grains in water (Brown, 1828)[1]. Albert Einstein gave a mathematical explanation of this phenomenon (Einstein, 1905). A rigorous mathematical construction of the random process known as Brownian motion or Wiener process was made by Norbert Wiener (1923). The work was motivated by Physics; however, as we discussed in Chapter 1, as early as 1900, Bachelier (1900) proposed Brownian motion as a model for stock price behaviour.

17.1 Symmetric Random Walk

To prepare ourselves for the introduction of the Wiener process in the next section, we look at some properties of the symmetric random walk in this section.

The stochastic process η_n, $n = 0, \ldots, N$, is a *symmetric random walk* under the measure P if $\eta_0 = 0$ and $\eta_n = \sum_{i=1}^{n} \zeta_i$, where the ζ_i can take only the values $\{-1, 1\}$ and are i.i.d. random variables under the measure P with

$$P(\zeta_i = \pm 1) = \frac{1}{2}.$$

This process is often motivated as a model of the gains from repeated plays of a fair game. For example, suppose we play a game in which each play is equivalent to tossing a fair coin. If it comes up heads I pay you £1, otherwise you pay me £1. For each n, the η_n is equal to my net gain after n plays. It is easy to compute (please do these calculations yourself) that

$$E_P \eta_n = 0 \ \text{ and } \ Var_P \eta_n = n.$$

---

[1]You can read a paper by Brian J. Ford on Brown's experiment (Ford, 1992) and see its recreation at the website http://www.sciences.demon.co.uk/wbbrowna.htm.

The random variables

$$\Delta\eta_n := \eta_n - \eta_{n-1}, \quad n = 1, 2, \ldots,$$

are called *increments* of the random walk. We see that $\Delta\eta_n = \zeta_n$ and thus the increments $\Delta\eta_n$ are independent under P. We say that this process is a *process with independent increments*.

The *quadratic variation* of a discrete-time stochastic process η_k up to time n is defined as

$$[\eta, \eta]_n := \sum_{i=1}^{n} (\Delta\eta_i)^2.$$

We emphasise that this quantity is computed path-by-path. It is not difficult to see that for the symmetric random walk $[\eta, \eta]_n = n$. We observe that $[\eta, \eta]_n$ coincides with $Var_P\eta_n$ but the natures of these two quantities are different. The variance is computed by averaging over all paths and it depends on the probability, whilst the quadratic variation is computed along a single path and is independent of our choice of probability. For a random walk, the quadratic variation $[\eta, \eta]_n$ does not depend on the particular path but, as a rule, the quadratic variation of a stochastic process does depend on the path, i.e. in general it is random.

Properties of the symmetric random walk

1. η_n is a martingale under the measure P.
2. η_n is a random process with independent increments.
3. The quadratic variation $[\eta, \eta]_n = n$, which is independent of the path along which it is computed.

Recall that the first property was proved in Problem 13 from Chapter 13.

Let us study the scaling properties of this random walk. Consider a time interval $[0, T]$ and divide it into N equal parts: $\Delta t = T/N$ and $t_k = k\Delta t$, $k = 0, 1, \ldots$. Introduce the scaled random walk $\eta_N(t_k)$:

$$\eta_N(0) = 0, \tag{17.1}$$

$$\eta_N(t_k) = \eta_N(t_{k-1}) + \sqrt{\Delta t}\,\zeta_k, \quad k = 1, \ldots, N,$$

with ζ_k the same as before. Note that the scaling has been done here in the manner consistent with our scaling in the binomial price model (in both cases the variance is proportional to time, cf. (16.5) and (17.4) below).

Obviously, the scaled random walk (17.1) also has properties 1 and 2 stated above, and, in particular, the increments

$$\Delta_k\eta_N := \eta_N(t_k) - \eta_N(t_{k-1}), \quad k = 1, 2, \ldots,$$

are independent. The quadratic variation of $\eta_N(t_k)$ is equal to

$$[\eta_N, \eta_N](t_k) = t_k.$$

We can complete the definition (17.1) of the scaled random walk by assuming that for $t_{k-1} < t < t_k$ we define $\eta_N(t)$ by linear interpolation between its values at the nearest points t_{k-1} and t_k, i.e. we connect the consecutive values $\eta_N(t_{k-1})$ and $\eta_N(t_k)$ by line segments:

$$\eta_N(t) := \eta_N(t_{k-1})\frac{t_k - t}{\Delta t} + \eta_N(t_k)\frac{t - t_{k-1}}{\Delta t} \quad \text{for } t_{k-1} < t < t_k. \quad (17.2)$$

We have

$$E(\eta_N(t) - \eta_N(s)) = 0, \quad (17.3)$$

$$Var(\eta_N(t_k) - \eta_N(t_l)) = t_k - t_l, \quad t_k \geq t_l.$$

In particular,

$$E\eta_N(t) = 0, \quad Var(\eta_N(t_k)) = t_k. \quad (17.4)$$

Exercise 17.1. Check the expressions (17.3) and (17.4).

What happens if we tend $\Delta t \to 0$? We should get an infinitesimal random walk. Let us consider the *limit properties of the scaled symmetric random walk*. First, we answer the question: what happens to the value $\eta_N(t)$ for a fixed $t > 0$ when $\Delta t \to 0$.

We have for $t_{k-1} \leq t \leq t_k$:

$$\eta_N(t) \stackrel{(17.2)}{=} \eta_N(t_{k-1})\frac{t_k - t}{\Delta t} + \eta_N(t_k)\frac{t - t_{k-1}}{\Delta t}$$

$$= \eta_N(t_k) + [\eta_N(t_{k-1}) - \eta_N(t_k)]\frac{t_k - t}{\Delta t}$$

$$\stackrel{(17.1)}{=} \sqrt{\Delta t}\sum_{i=1}^{k}\zeta_i - \zeta_k\frac{t_k - t}{\sqrt{\Delta t}} = \sqrt{t}\sqrt{\frac{t_k}{t}}\frac{\sum_{i=1}^{k}\zeta_i}{\sqrt{k}} - \zeta_k\frac{t_k - t}{\sqrt{\Delta t}}.$$

We observe that as $\Delta t \to 0$

$$\frac{t_k}{t} \to 1,$$

$$0 \leq \frac{t_k - t}{\sqrt{\Delta t}} \leq \frac{\Delta t}{\sqrt{\Delta t}} = \sqrt{\Delta t} \to 0.$$

Due to CLT, we have

$$\frac{\sum_{i=1}^{k}\zeta_i}{\sqrt{k}} \Rightarrow \tilde{\theta} \sim \mathcal{N}(0, 1) \quad \text{as } \Delta t \to 0.$$

Summarising: we have obtained that for each fixed t the distribution of the scaled random walk $\eta_N(t)$ evaluated at time t converges to the normal distribution with zero mean and variance t:

$$\eta_N(t) \Rightarrow \sqrt{t}\tilde{\theta} \sim \mathcal{N}(0, t).$$

Another question we might ask is what happens to the increment $\eta_N(t) - \eta_N(s)$, $t > s$, as $\Delta t \to 0$? A similar reasoning implies the convergence in distribution

$$\eta_N(t) - \eta_N(s) \Rightarrow \mathcal{N}(0, |t - s|) \quad \text{as} \quad \Delta t \to 0.$$

Exercise 17.2. Show that the above is true.

It is reasonable to think that the properties of the scaled random walk can be preserved for an infinitesimal random walk and hence we naturally come to the next section where we define the corresponding continuous-time stochastic process.

17.2   Wiener Process

We note that Definition 9.1 – the definition of a stochastic process – works both in the discrete and continuous case.

We define Wiener process (Brownian motion), which can be viewed as a limit (in some sense, which is not specified here, but also see Remark 17.1) of the scaled symmetric random walk, in an axiomatic way.

Definition 17.1. *Standard* **Wiener process** $W(t)$, $t \geq 0$, *is a random process such that*

1. $W(0) = 0$.
2. *For any $0 \leq t_0 < t_1 < \cdots < t_n$, the increments $W(t_1) - W(t_0), W(t_2) - W(t_1), \ldots, W(t_n) - W(t_{n-1})$ are independent.*
3. *The random variable $W(t) - W(s)$, $s < t$, has the normal distribution with zero mean, $E(W(t) - W(s)) = 0$, and variance $Var(W(t) - W(s)) = t - s$, i.e. the distribution function of this random variable is such that*

$$P(a < W(t) - W(s) < b) = \frac{1}{\sqrt{2\pi(t-s)}} \int_a^b \exp\left(-\frac{z^2}{2(t-s)}\right) dz.$$

4. *The trajectories of $W(t)$ are continuous.*

There are different proofs for the existence of this mathematical object[2]. We will not consider them.

Remark 17.1. By the multidimensional CLT (see Chapter VII in Shiryaev, 1996), one can show that for all $t_1, \ldots, t_k$ the finite dimensional distribution $F_{\eta_N(t_1), \ldots, \eta_N(t_k)}(x_1, \ldots, x_k)$ of the scaled random walk (17.1), (17.2) converges to the finite-dimensional distribution $F_{W(t_1), \ldots, W(t_k)}(x_1, \ldots, x_k)$ as $\Delta t \to 0$, where $W(t)$ is a standard Wiener process[3]. One can prove even more that the distribution of $\eta_N(t)$, $t \geq 0$, converges (in some sense) to the distribution of $W(t)$, $t \geq 0$[4].

17.3   Properties of the Wiener process

From the second property in Definition 17.1 we see that, like the simple random walk, the Wiener process is a process with independent increments. Consequently, the joint density of Wiener increments has the form

$$
\rho(x_1, \ldots, x_n) = \prod_{k=1}^{n} \frac{1}{\sqrt{2\pi(t_k - t_{k-1})}} \exp\left(-\frac{x_k^2}{2(t_k - t_{k-1})}\right)
$$
$$
= \frac{1}{(2\pi)^{\frac{n}{2}} \prod_{k=1}^{n} \sqrt{t_k - t_{k-1}}} \exp\left\{-\sum_{k=1}^{n} \frac{x_k^2}{2(t_k - t_{k-1})}\right\}.
$$

The Wiener process also has the following properties.

1. $EW(t) = 0$, $E[W(t)]^2 = t$.
2. Trajectories of $W(t)$ are not differentiable (almost everywhere).
3. *Quadratic variation.* Let $t_0 < \cdots < t_N = T$ be an arbitrary discretization on an interval $[t_0, T]$ into N parts and $h := \max_k(t_{k+1} - t_k)$. Introduce the quadratic variation of the Wiener process:

$$
\vartheta_h := \sum_{k=0}^{N-1} [W(t_{k+1}) - W(t_k)]^2.
$$

Then

$$
E(\vartheta_h - (T - t_0))^2 \to 0 \quad \text{as} \quad h \to 0. \tag{17.5}
$$

The limit we have in (17.5) is called *mean-square convergence* and we say that ϑ_h converges to $T - t_0$ in the mean-square sense.

---

[2]See, e.g., Section 1.5 in Ito and McKean (1965) or Section 3.2 in Etheridge (2002).
[3]In Bachelier's analysis (Bachelier, 1900) a Brownian motion arose as a (formal) limit of a simple random walk.
[4]See, e.g., Billingsley (1968).

Example 17.1. We prove the second equality in property 1:

$$E[W(t)]^2 = E[W(t) - W(0)]^2 = t - 0 = t.$$

Exercise 17.3. Prove the first equality in property 1.

Heuristic argument for justification of property 2:
Define

$$\eta_h(t) := \frac{W(t+h) - W(t)}{h}.$$

Obviously, $\eta_h(t) \sim \mathcal{N}(0, 1/h)$. Hence $\eta_h(t)$ cannot have a limit as $h \to 0$ in any probabilistic sense. Note that although the Wiener process is continuous everywhere, it is (with probability 1) differentiable nowhere.

Example 17.2. (*Hint for the proof of the property about quadratic variation*). First, show that

$$E\vartheta_h = T - t_0 \ \text{ and } \ Var(\vartheta_h) = 2 \sum_{k=0}^{N-1} (t_{k+1} - t_k)^2. \tag{17.6}$$

Then

$$E(\vartheta_h - (T - t_0))^2 = E(\vartheta_h - E\vartheta_h)^2 = Var(\vartheta_h)$$
$$= 2 \sum_{k=0}^{N-1} (t_{k+1} - t_k)^2 \to 0 \ \text{ as } h \to 0.$$

Exercise 17.4. Derive (17.6) and thus complete the above proof.

Remark 17.2. Let us pay attention to the fact that the quadratic variation property of the Wiener process is rather unusual. For a smooth function, quadratic variation (i.e. sum of squares of increments) tends to zero as $h \to 0$ (*Exercise.* Check that this is right). For functions changing by some finite jumps, such a sum tends to the sum of squares of jumps on the interval $[t_0, T]$ and the limit does not depend on t_0, T in a continuous manner (*Exercise.* Check that this is right).

Remark 17.3. The convergence of the quadratic variation in the mean-square sense (17.5) (also called L_2-convergence) does not imply that for almost all trajectories of the Wiener process the limit

$$\lim_{h \to 0} \sum_{k=0}^{N-1} [W(t_{k+1}) - W(t_k)]^2$$

exists. The mean-square convergence implies that there is a subsequence of partitions of the interval $[t_0, T]$ along which the convergence to $T - t_0$ is almost sure (i.e. for almost all trajectories or, in other words, the convergence takes place for all trajectories except for a set of them having probability zero on which the assertion might be not true). See further details in, e.g., Karatzas and Shreve (1991); Wentzell (1981) and Rogers and Williams (2000).

Corollary to property 3. *For almost all trajectories of the Wiener process $W(t)$ the (first-order) variation $\sum_{k=0}^{N-1} |W(t_{k+1}) - W(t_k)|$ is unbounded on any interval.*

Proof. Let us consider a subsequence of partitions of the interval $[t_0, T]$ for which $\lim_{h \to 0} \sum_{k=0}^{N-1} [W(t_{k+1}) - W(t_k)]^2$ equals $T - t_0$ almost surely (see Remark 17.3). We have

$$\sum_{k=0}^{N-1} [W(t_{k+1}) - W(t_k)]^2 \leq \max_k |W(t_{k+1}) - W(t_k)| \times \sum_{k=0}^{N-1} |W(t_{k+1}) - W(t_k)|.$$
(17.7)

Due to the continuity of $W(t)$, $\max_k |W(t_{k+1}) - W(t_k)| \to 0$ as $h \to 0$. The quadratic variation in the left-hand side of (17.7) goes to the finite $T - t_0$ as $h \to 0$ (see Property 3 of the Wiener process and Remark 17.3). Then the variation $\sum_{k=0}^{N-1} |W(t_{k+1}) - W(t_k)| \to \infty$ as $h \to 0$ which proves the corollary.

The consequence of this result is absolutely astonishing. Variation basically gives you the length of the trajectory, therefore, a trajectory of the Wiener process has infinite length on any finite time interval!

Some other properties.

4. A trajectory of the Wiener process will eventually take any (whatever large or small) value.

5. At every scale a trajectory of the Wiener process looks the same. It is a fractal[5]. (In this respect it is useful to recall the observation about asset price behaviour made after Exercise 14.1 in Section 14.2).

A typical trajectory of the Wiener process is plotted in Fig. 17.1.

Thus we see that the Wiener process has rather unusual properties from the point of view of standard Analysis. Between 1940 and 1950, Calculus

---

[5]A geometric pattern that is repeated at even smaller scales to produce irregular shapes and surfaces that cannot be represented by classical Geometry.

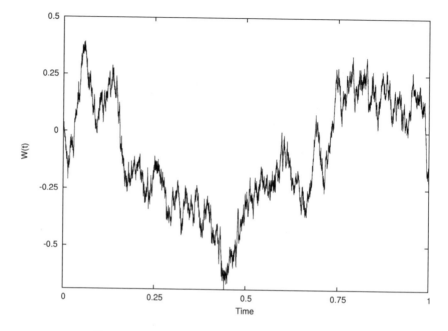

Fig. 17.1   A typical trajectory of the Wiener process.

based on the Wiener process arose. This Calculus we often now call *Ito Calculus* and the theory of stochastic differential equations. This theory was created independently by K. Ito (Japan) and I. I. Gichman (USSR) during the Second World War. They continued the study undertaken by S. N. Bernstein (USSR) and N. Wiener (USA).

17.4   Geometric Brownian Motion

Let $T = N\Delta t$. Consider again the discrete model (I) from Chapter 7:

$$S_n = S_{n-1} + S_{n-1}\xi_n, \quad n = 0, 1, \ldots, N, \tag{Ia}$$

where ξ_n are random variables which can take one of two values (as we argued before, see (16.4) in Chapter 16):

$$b = \exp(\mu\Delta t + \sigma\sqrt{\Delta t}) - 1, \quad a = \exp(\mu\Delta t - \sigma\sqrt{\Delta t}) - 1$$

with either

1. the 'actual' (market) probabilities

$$P(\xi_n = a) = P(\xi_n = b) = \frac{1}{2};$$

or

2. the risk-neutral probabilities

$$P(\xi_n = b) = q = \frac{e^{r\Delta t} - (1 + a)}{b - a}, \quad P(\xi_n = a) = 1 - q.$$

Recall that when we are pricing derivatives, we are using the risk-neutral probabilities.

Let us take any

$$t = L\Delta t \le T.$$

We know that in case 1 (see, e.g., (16.8) in Section 16.1):

$$\ln \frac{S_L}{S_0} = \mu t + \sigma \sqrt{\Delta t} \sum_{k=1}^{L} \vartheta_k, \tag{17.8}$$

where ϑ_k are random variables taking the values ± 1 with probability $1/2$. The sum $\sqrt{\Delta t} \sum_{k=1}^{L} \vartheta_k$ is a scaled random walk $\eta_N(t)$ (see Section 17.1) and we can write

$$\ln \frac{S_L}{S_0} = \mu t + \sigma \eta_N(t) \quad \text{for } t = L\Delta t. \tag{17.9}$$

Further, we argued before (see Sections 17.1 and 17.2) that the scaled random walk converges (in some sense) to the Wiener process as $\Delta t \to 0$. Then we can say that as $\Delta t \to 0$

$$\ln \frac{S_L}{S_0} \to \ln \frac{S(t)}{S_0} = \mu t + \sigma W(t), \tag{17.10}$$

where $W(t)$ is a standard Wiener process from Definition 17.1. Hence, we (not rigorously, just heuristically) arrive at the *continuous price model*

$$S(t) = S_0 e^{\mu t + \sigma W(t)}, \tag{17.11}$$

where $W(t)$ is a standard Wiener process.

Definition 17.2. *The stochastic process defined by (17.11) is called **geometric Brownian motion**.*

Geometric Brownian motion is the basic reference model for stock prices in continuous time. In particular, it is the model underlying the Black–Scholes–Merton pricing theory.

To price options, as we know, we need to work with risk-neutral probabilities – case 2 above. If we would take a heuristic consideration of the limit $\Delta t \to 0$, similar to how we have done this above in the case of market probabilities, we obtain the following process:

$$S(t) = S_0 e^{(r - \frac{\sigma^2}{2})t + \sigma W(t)}. \tag{17.12}$$

Again, this is geometric Brownian motion.

Remark 17.4. As in the discrete case we considered before in detail, (17.11) and (17.12) are understood as stock price models written under two different measures and we can come from (17.11) to (17.12) by changing measure from 'real' to 'risk-neutral', i.e. from P to Q^6. In particular, a process may be a Wiener process under the probability measure P, but the same process is not a Wiener process under a different measure Q (i.e. $W(t)$ in (17.11) and (17.12) are different: $W(t)$ in (17.11) is a Wiener process under the market measure while $W(t)$ in (17.12) is a Wiener process under the risk-neutral measure). As in the discrete case, we divorce stochastic processes and probability measures.

17.5   Basics of Continuous-Time Stochastic Processes

Recall that working with continuous random variables (probabilistic model for experiments with infinitely many outcomes) we assume that we have the corresponding probability space $(\Omega, \mathcal{F}, P)$ which is sufficiently rich for our experiment. Here, Ω is the sample space, $\mathcal{F}$ – a σ-algebra of events and P – a probability measure. Let us revisit some notions considered in Chapter 9 for discrete stochastic processes and introduce them in the continuous-time case. We start with the definition of filtration (compare with Definition 9.1).

Definition 17.3. *Filtration* $\{\mathcal{F}_t\}_{t \geq 0}$ *is a set of σ-algebras $\mathcal{F}_t$ such that* $\mathcal{F}_s \subseteq \mathcal{F}_t$ *for $s < t$ and $\mathcal{F}_t \subseteq \mathcal{F}$ for all t.*

As in the discrete case, *natural filtration* $\{\mathcal{F}_t^X\}_{t \geq 0}$ associated with a stochastic process $X(t)$ encodes the information generated by this process on the interval $[0, t]$. We say that an event $A \in \mathcal{F}_t^X$ if, having observed the

---

[6]The tool for doing this is Girsanov's theorem (see, e.g., Etheridge, 2002; Björk, 2004; Lamberton and Lapeyre, 2007; Shiryaev, 1999).

trajectory $X(s)$ for $s \in [0, t]$, we can answer the question of whether or not A has occurred. Obviously, more than one process can be associated with the same filtration. In what follows, we will usually associate filtration $\mathcal{F}_t$ with the Wiener process $W(t)$ so that

1. For all $t \geq 0$ the random variable $W(t) = W(t, \omega)$ is measurable with respect to $\mathcal{F}_t$.
2. The increment $W(t+s) - W(t)$ for all $s \geq 0$ does not depend on $\mathcal{F}_t$ which means that the random variable $W(t + s) - W(t)$ is independent of all random variables $W(s')$, $0 \leq s' \leq t$.

Definition 17.4. *We say that a stochastic process $X(t)$ is **adapted** to the filtration $\{\mathcal{F}_t\}_{t \geq 0}$ if $X(t)$ is measurable with respect to $\mathcal{F}_t$ for all $t \geq 0$.*

Example 17.3. The process $Y(t) := \max_{0 \leq s \leq t} W(s)$ is adapted to the filtration $\{\mathcal{F}_t^W\}_{t \geq 0}$ generated by the Wiener process $W(t)$, whereas the process $Z(t) := W^2(t + 1) - W^2(t)$ is not adapted to $\{\mathcal{F}_t^W\}_{t \geq 0}$.

Definition 17.5. *The quadruple $(\Omega, \mathcal{F}, \{\mathcal{F}_t\}_{t \geq 0}, P)$ is called a **filtered probability space**.*

Chapter 18

Simplistic Introduction to Ito Calculus

'It is useful to solve differential equations.'[1]

First, consider a bank account, for which we use the model

$$B(t) = B_0 e^{rt}, \tag{18.1}$$

where $r \geq 0$ is a constant interest rate. Equivalently, we can say that $B(t)$ is the solution of the following initial value problem for the ordinary differential equation (ODE):

$$\dot{B} = rB, \quad B(0) = B_0, \tag{18.2}$$

or

$$dB = rBdt, \quad B(0) = B_0,$$

or in the integral form

$$B(t) = B_0 + \int_0^t rB(s)ds.$$

As we discussed in Chapter 4, in real life interest rates are not constant and (18.2) can be generalised by substituting the constant r with a short rate depending on time $r(t)$ and possibly on randomness $r(t) = r(t, \omega)$ and some other variables:

$$\dot{B} = r(t)B, \quad B(0) = B_0. \tag{18.3}$$

Newton saw the usefulness of ODEs because by using them one can describe laws of nature. Now let us try to answer the question: what kind of differential equation can we write for geometric Brownian motion

$$S(t) = S_0 e^{\mu t + \sigma W(t)}? \tag{18.4}$$

---

[1]V. I. Arnold's (1983) interpretation of Sir Isaac Newton's anagram contained in a letter to Gottfried Leibniz in 1677.

Take a 'nice' (smooth) function $g(t, x)$ of two independent ('dummy') variables. Due to the Taylor theorem, we have

$$g(t + \Delta t, x + \Delta x) - g(t, x) = \frac{\partial g}{\partial t} \Delta t + \frac{\partial g}{\partial x} \Delta x + \frac{1}{2} \frac{\partial^2 g}{\partial x^2} (\Delta x)^2 + \cdots. \quad (18.5)$$

If in (18.5) we substitute a smooth function $x(t)$ instead of the dummy variable x and $\Delta x = x(t + \Delta t) - x(t)$, the last term of the right-hand side of (18.5) is of order[2] $O((\Delta t)^2)$ and the omitted terms in (18.5) are of order $O((\Delta t)^2)$ or higher. Then, as $\Delta t \to 0$, we get the differential[3]

$$dg = \frac{\partial g}{\partial t} dt + \frac{\partial g}{\partial x} dx$$

or

$$\frac{dg}{dt}(t, x(t)) = \frac{\partial g}{\partial t} + \frac{\partial g}{\partial x} \frac{dx}{dt},$$

which is just the chain rule.

Let

$$g(t, x) = S_0 e^{\mu t + \sigma x}. \quad (18.6)$$

Then we can write $S(t) = g(t, W(t))$ and obtain

$$g(t + \Delta t, W(t + \Delta t)) - g(t, W(t)) \quad (18.7)$$

$$= \frac{\partial g}{\partial t} \Delta t + \frac{\partial g}{\partial x} (W(t + \Delta t) - W(t)) + \frac{1}{2} \frac{\partial^2 g}{\partial x^2} (W(t + \Delta t) - W(t))^2 + \cdots.$$

We know (see Definition 17.1) that

$$E[W(t + \Delta t) - W(t)]^2 = \Delta t.$$

Hence, the last term on the right-hand side of (18.7) is, in fact, of order Δt and we cannot ignore it, i.e. we cannot ignore the term containing the second derivative as we did above in the case of a smooth $x(t)$. Taking this into account and assuming that we can do the transition in (18.7) as $\Delta t \to 0$ rigorously, the differential equation governing $S_t = g(t, W_t)$ will take the form

$$dS = \frac{\partial g}{\partial t}(t, W(t))dt + \frac{\partial g}{\partial x}(t, W(t))dW(t) + \frac{1}{2} \frac{\partial^2 g}{\partial x^2}(t, W(t))dt. \quad (18.8)$$

Due to (18.6), we have

$$\frac{\partial g}{\partial t} = S_0 \mu e^{\mu t + \sigma x} = \mu g, \quad \frac{\partial g}{\partial x} = S_0 \sigma e^{\mu t + \sigma x} = \sigma g, \quad \frac{\partial^2 g}{\partial x^2} = \sigma^2 g.$$

---

[2] Please check.

[3] The differential of a function is the principal (linear) part in the increment of a function.

Thus, we heuristically get the **stochastic differential equation (SDE)** for geometric Brownian motion:

$$dS = (\mu + \frac{\sigma^2}{2})S dt + \sigma S dW(t), \ S(0) = S_0, \tag{18.9}$$

or

$$\frac{dS}{dt} = (\mu + \frac{\sigma^2}{2})S + \sigma S \dot{W}(t), \ S(0) = S_0. \tag{18.10}$$

However, we still have a major problem here: how to interpret $dW(t)$ given that $W(t)$ is a 'bad' function (it is not differentiable, its variation is unbounded). Due to this, (18.9) actually has no sense in the form it is written[4]. It is just a formal writing of a mathematical object. There are at least two ways to interpret it. One is simple but less productive, the other is more productive but less simple.

The first way
Consider an ODE:

$$\dot{x} = g(t, x), \ x(0) = x_0,$$

and forget for the moment that you know what derivative is and that the definition of this object is based on the notion of derivatives. Then we can view this ODE as a limiting object of the (well-defined) difference equation (the Euler method[5]):

$$x_{k+1} = x_k + \delta g(t_k, x_k)$$

with a discretization step $\delta = T/N$ of the time interval $[0, T]$ with $0 = t_0 < t_1 < \cdots < t_N = T$. If the limit of x_k exists when $\delta \to 0$, we can say that the ODE is such an object for which solution $x(t)$ is the obtained limit.

We can do something similar with the stochastic differential equation (18.9) – we can view it as a limit of one or another discrete scheme, e.g., of the (weak) Euler method[6]:

$$S_{k+1} = S_k + (\mu + \frac{\sigma^2}{2})S_k \delta + \sigma S_k \sqrt{\delta} \xi_{k+1}, \tag{18.11}$$

where ξ_k are i.i.d. with $P(\xi_k = \pm 1) = 1/2$ or we can take ξ_k as i.i.d. Gaussian random variables with zero mean and unit variance (this may ensure

---

[4]More precisely, it is so if we look at it from the position of standard Analysis. The modern theory of white noise (see, e.g., Holden *et al.* (2010) and references therein) gives a rigorous interpretation to (18.10) but it is far beyond the scope of this course and it is also not used in Financial Engineering.

[5]See, e.g., Cheney and Kincaid (2012).

[6]See, e.g., Glasserman (2003) and Milstein and Tretyakov (2004).

a stronger convergence). The scheme (18.11) converges in some sense to a continuous process $S(t)$ which can be viewed as a solution of (18.9). This approach can be justified rigorously (it was first considered by S. N. Bernstein in the 1930s, (Bernstein, 1934)). We note that an approach similar to this was exploited by us on the heuristic level in Chapter 17, when we tended the scaled symmetric random walk to the limit.

The second way

This is more complicated. It is based on *Stochastic Calculus*. Due to this, a rigorous sense may be given to (18.9) if we rewrite it in the integral form:

$$S(t) = S_0 + \int_0^t (\mu + \frac{\sigma^2}{2})S(s)ds + \int_0^t \sigma S(s)dW(s). \tag{18.12}$$

Then, the equation (18.9) is used as the symbolic writing of (18.12). The first integral in (18.12) is just a usual Riemann integral because $S(s)$ is continuous. However, the second integral requires a rigorous mathematical interpretation. It is the so-called **Ito stochastic integral**. It was introduced by K. Ito (1942).

Before we go further in mathematics and define the Ito integral, it is best to understand a little bit about the use of SDE in Finance. The geometric Brownian motion is a very simplistic model for price behaviour. It assumes that, e.g., (log) volatility σ is constant. This cannot be true in practice (recall, for instance, the volatility smile effect; see Section 16.3 and also Björk, 2004; Musiela and Rutkowski, 2005; Shreve, 2004). It may depend on time, on price, or some other parameters. The introduction of differential equation (18.9) (i.e. (18.12)) allows us to generalise the geometrical Brownian motion model. For instance, we can write the following quite general equation as a model for stock price:

$$dS = a(t, S)dt + b(t, S)dW(t), \tag{18.13}$$

where $a(t, x)$ and $b(t, x)$ are some functions (possibly random). In other words, SDEs are useful for modelling purposes in the stochastic framework as ODEs are in the deterministic case.

18.1   Ito Integral and Stochastic Differential Equation

To complete the definition of (18.12) and (18.13), we need to give the meaning to an expression such that $\int_0^T h(t)dW(t)$, where $h(t) = h(t, \omega)$ is $\mathcal{F}_t$-measurable function for all $t > 0$.

Example 18.1. How is it possible to give the meaning to the integrals $\int_0^T W(t)dW(t)$ or $\int_0^T S(t)dW(t)$?

Assume for a moment that $W(t)$ is differentiable. Then, obviously

$$dW(t) = \dot{W}(t)dt, \quad W(0) = 0,$$

and for 'good' functions h

$$\int_0^T h(t)dW(t) = \int_0^T h(t)\dot{W}(t)dt,$$

where the integral on the right would be a well-defined Riemann integral. However, $W(t)$ is not differentiable as we know. If variation of $W(t)$ would be bounded, this integral can be interpreted as the Stieltjes (Riemann–Stieltjes) integral[7]. However, variation of $W(t)$ is unbounded. So, we need to do something else.

Let $h(t)$ be a continuous stochastic process adapted to $\mathcal{F}_t$ such that

$$\int_0^T E[h(t)]^2 dt < \infty.$$

Then the **Ito stochastic integral** is defined as the mean-square limit:

$$\int_0^T h(t)dW(t) := \operatorname*{l.i.m.}_{h \to 0} \sum_{i=0}^{N-1} h(t_i) \times (W(t_{i+1}) - W(t_i)), \qquad (18.14)$$

where $0 = t_0 < t_1 < \cdots < t_N = T$ is a discretization of the time interval $[0, T]$, $\delta := \max_i(t_{i+1} - t_i)$ and l.i.m. means the limit in the mean-square sense, i.e.

$$E\left[\int_0^T h(t)dW(t) - \sum_{i=0}^{N-1} h(t_i) \times (W(t_{i+1}) - W(t_i))\right]^2 \to 0 \quad \text{as } \delta \to 0.$$
$$(18.15)$$

Remark 18.1. Ito integral exists for any continuous, bounded (and some unbounded) $h(t)$ (see a proof in, e.g., Wentzell, 1981; Karatzas and Shreve, 1991; Krylov, 2002; Rogers and Williams, 2000).

---

[7]The Riemann–Stieltjes integral of a continuous function $g(t)$ with respect to a function $f(x)$ of bounded variation on $[0, T]$ is defined as

$$\int_0^T g(t)df(t) := \lim_{h \to 0} \sum_{i=0}^{N-1} g(s_i^*) \times (f(t_{i+1}) - f(t_i)),$$

where s_i^* is an arbitrary point in the interval $[t_i, t_{i+1}]$.

Remark 18.2. It is essential that $h(t_i)$ in (18.14) must always be evaluated at the left point, in contrast to the Riemann–Stieltjes integral (see Footnote 7) which does not depend on the point $t \in [t_i, t_{i+1}]$ at which h is evaluated. If we allow $h(t)$ to be evaluated at a different point in (18.14), the limit will be different and we will get another type of stochastic integral (see also Exercise 18.1). This is again unusual from the point of view of standard Analysis and is a consequence of, in particular, unbounded variation of the Wiener process.

Example 18.2. Compute the integral $\int_0^T W(t)dW(t)$.

Answer. We have for a discretization $0 = t_0 < t_1 < \cdots < t_N = T$:

$$\sum_{i=0}^{N-1} W(t_i)(W(t_{i+1}) - W(t_i))$$

$$= \sum_{i=0}^{N-1} \left\{ \frac{1}{2}W^2(t_{i+1}) - \frac{1}{2}W^2(t_{i+1}) + W(t_i)W(t_{i+1}) - W^2(t_i) \right\}$$

$$= \sum_{i=0}^{N-1} \left\{ \frac{1}{2}[W^2(t_{i+1}) - W^2(t_i)] - \frac{1}{2}(W(t_{i+1}) - W(t_i))^2 \right\}$$

$$= \frac{1}{2}\sum_{i=0}^{N-1}[W^2(t_{i+1}) - W^2(t_i)] - \frac{1}{2}\sum_{i=0}^{N-1}[W(t_{i+1}) - W(t_i)]^2$$

$$= \frac{1}{2}W^2(T) - \frac{1}{2}\sum_{i=0}^{N-1}[W(t_{i+1}) - W(t_i)]^2.$$

The second term in the last line of the above equation is a quadratic variation which, as we know (see (17.5)), tends to T as $\delta \to 0$. Then we have the following limit in the mean-square sense:

$$\sum_{i=0}^{N-1} W(t_i)(W(t_{i+1}) - W(t_i)) \to \frac{W^2(T)}{2} - \frac{T}{2} \quad \text{as} \ \ \delta \to 0$$

and, consequently,

$$\int_0^T W(t)dW(t) = \frac{W^2(T)}{2} - \frac{T}{2}.$$

This is not what one would predict from the classical integration theory. The extra term $-T/2$ arises due to the fact that the quadratic variation of the Wiener process is not zero.

Exercise 18.1. Compute l.i.m.$_{h\to 0}\sum_{i=0}^{N-1} W(t_{i+1})(W(t_{i+1}) - W(t_i))$ and compare with the answer from Example 18.2 (also compare with Remark 18.2).

Properties of Ito integral

1. $E \displaystyle\int_0^T h(t)dW(t) = 0;$

2. $E \left[\displaystyle\int_0^T h(t)dW(t)\right]^2 = \displaystyle\int_0^T E[h(t)]^2 dt.$

Proofs of these properties can be found, e.g., in Krylov (2002) and Rogers and Williams (2000).

Remark 18.3. If $h(t)$ does have a continuous first derivative (note that the integrands in Example 18.1 are not differentiable), then, e.g.,

$$\underset{h\to 0}{\text{l.i.m.}} \sum_{i=0}^{N-1} h(t_i) \times (W(t_{i+1}) - W(t_i)) \tag{18.16}$$

$$= \underset{h\to 0}{\text{l.i.m.}} \sum_{i=0}^{N-1} h(t_{i+1}) \times (W(t_{i+1}) - W(t_i))$$

or, more precisely, the definition of the stochastic integral in this case does not depend on the point $t \in [t_i, t_{i+1}]$ at which h is evaluated (cf. Footnote 7). Further, one can show that for differentiable $h(t)$:

$$\int_0^T h(t)dW(t) = h(T)dW(T) - \int_0^T W(t)h'(t)dt. \tag{18.17}$$

Exercise 18.2. (i) Show that (18.16) is true. (ii) Show that (18.17) is true.

Based on the definition of the Ito integral, we can now properly define SDE. Assume that $a(t, x)$ and $b(t, x)$ have some good analytical properties. The solution of the stochastic differential equation

$$dX = a(t, X)dt + b(t, X)dW(t), \quad X(t_0) = x, \tag{18.18}$$

is a stochastic process $X(t)$ such that for every t the random variable $X(t)$ is measurable with respect to $\mathcal{F}_t$ and $X(t)$ satisfies the integral equation

$$X(t) = x + \int_{t_0}^t a(s, X(s))ds + \int_{t_0}^t b(s, X(s))dW(s), \tag{18.19}$$

where the first integral is a Riemann integral and the second one is an Ito integral. A proof of the existence and uniqueness of solutions of SDEs under some assumptions on the functions $a(t, x)$ and $b(t, x)$ can be found in, e.g., Karatzas and Shreve (1991); Wentzell (1981); Hasminskii (1980); Krylov (2002) and Rogers and Williams (2000).

18.2　Ito Formula

In this section we consider the Ito formula which is one of the most important results in Stochastic Calculus. Here we give its heuristic justification. Its rigorous proof can be found in, e.g., Karatzas and Shreve (1991); Wentzell (1981); Krylov (2002) and Rogers and Williams (2000).

First, we recall (see also the beginning of Chapter 18) that if $g(t, x)$, $x \in \mathbf{R}$, is a smooth deterministic function and $X(t)$ is a solution of the ODE

$$dX = a(t, X(t))dt,$$

then the differential of $g(t, X(t))$ is found as

$$dg(t, X(t)) = \left[\frac{\partial g}{\partial t}(t, X(t)) + a(t, X(t)) \frac{\partial g}{\partial x}(t, X(t)) \right] dt. \qquad (18.20)$$

Consider the (one-dimensional) SDE

$$dX = a(t, X)dt + b(t, X)dW(t), \quad X(t_0) = x. \qquad (18.21)$$

Let us find a formula for $g(t, X(t))$ with $X(t)$ from (18.21) analogous to (18.20).

Following the Taylor theorem, we formally have (see also (18.5) and (18.7)):

$$dg(t, X(t)) = \frac{\partial g}{\partial t}(t, X)dt + \frac{\partial g}{\partial x}(t, X)dX \qquad (18.22)$$

$$+ \frac{1}{2}\frac{\partial^2 g}{\partial x^2}(t, X) \times (dX)^2 + \cdots$$

$$= \frac{\partial g}{\partial t}(t, X)dt + \frac{\partial g}{\partial x}(t, X)\, a(t, X)dt + \frac{\partial g}{\partial x}(t, X)\, b(t, X)dW(t)$$

$$+ \frac{1}{2}\frac{\partial^2 g}{\partial x^2}(t, X)\, b^2(t, X)\, (dW)^2 + \cdots,$$

where the omitted terms contain factors of the form $(dW)^3$, $(dW)^2 dt$, $(dt)^2$, $(dt)^2 dW$, etc. Mean-square orders of smallness of all these terms are higher than dt. We already know that $E[dW(t)]^2 = dt$ and that the last term in (18.22) cannot be omitted: it becomes, after integration,

$$\frac{1}{2}\int_{t_0}^{t} \frac{\partial^2 g}{\partial x^2}(s, X)\, b^2(s, X)ds.$$

Finally, integrating (18.22), we obtain (we assume here that we can do the transition rigorously):

$$g(t, X(t)) - g(t_0, x) = \int_{t_0}^t \left[\frac{\partial g}{\partial t}(s, X(s)) + a(s, X(s))\frac{\partial g}{\partial x}(s, X(s)) \right.$$

$$\left. + \frac{b^2(s, X(s))}{2}\frac{\partial^2 g}{\partial x^2}(s, X(s)) \right] ds \qquad (18.23)$$

$$+ \int_{t_0}^t b(s, X(s))\frac{\partial g}{\partial x}(s, X(s))\, dW(s)$$

or in the form of differentials

$$dg(t, X(t)) = \left[\frac{\partial g}{\partial t}(t, X(t)) + a(t, X(t))\frac{\partial g}{\partial x}(t, X(t)) \right. \qquad (18.24)$$

$$\left. + \frac{b^2(t, X(t))}{2}\frac{\partial^2 g}{\partial x^2}(t, X(t)) \right] dt$$

$$+ b(t, X(t))\frac{\partial g}{\partial x}(t, X(t))\, dW(t),$$

which is the **Ito formula**. It is also known as the formula of the change of variables in the stochastic integral.

Example 18.3. Let $g(x) = x^2$ and $S(t)$ be a solution of (18.12). Find the SDE for $\xi(t) := g(S(t))$ and then compute $ES^2(t)$.

Answer. Applying the Ito formula, we obtain

$$dg(S(t)) = \left[(\mu + \frac{\sigma^2}{2})S(t)\frac{\partial g}{\partial x}(S(t)) + \frac{\sigma^2}{2}S^2(t)\frac{\partial^2 g}{\partial x^2}(S(t)) \right] dt$$

$$+ \sigma S(t)\frac{\partial g}{\partial x}(S(t))\, dW(t)$$

$$= 2\left[\mu + \sigma^2 \right] S^2(t)dt + 2\sigma S^2(t)\, dW(t),$$

i.e.

$$d\xi = 2\left[\mu + \sigma^2 \right]\xi dt + 2\sigma\xi dW(t), \quad \xi(0) = S^2(0).$$

Formally taking expectations from the left- and right-hand sides of the above equation and taking into account that the expectation of the Ito integral is zero, we obtain

$$dE\xi = 2\left[\mu + \sigma^2 \right] E\xi dt, \quad E\xi(0) = S^2(0).$$

Hence

$$E\xi(t) = \xi(0)\exp(2\left[\mu + \sigma^2 \right] t),$$

i.e.

$$ES^2(t) = S^2(0) \exp(2 \left[\mu + \sigma^2 \right] t).$$

Example 18.4. (*Probabilistic representation of solutions of parabolic partial differential equations*). Define the parabolic linear differential operator L:

$$L := \frac{\partial}{\partial t} + a(t, x)\frac{\partial}{\partial x} + \frac{b^2(t, x)}{2}\frac{\partial^2}{\partial x^2}, \quad \sigma \geq \sigma_0 > 0. \tag{18.25}$$

Consider the Cauchy problem for the parabolic partial differential equation (PDE):

$$Lu = 0, \quad t \in [t_0, T), \quad x \in \mathbf{R}, \tag{18.26}$$

with the terminal condition

$$u(T, x) = \varphi(x), \quad x \in \mathbf{R}, \tag{18.27}$$

together with the associated SDE

$$dX = a(s, X)ds + b(s, X)dW(s), \quad X(t) = x, \quad s \geq t. \tag{18.28}$$

We assume that there exists a unique sufficiently smooth 'good' function $u(t, x)$ satisfying (18.26)–(18.27), i.e. we assume the existence and uniqueness of a sufficiently smooth solution of the Cauchy problem (18.26)–(18.27)[8]. We apply the Ito formula to $u(s, X_{t,x}(s))$, where, $X_{t,x}(s)$, $s \geq t$, is the solution of the SDE (18.28) with the initial condition $X(t) = x$. As a result, we obtain

$$u(T, X_{t,x}(T)) - u(t, x) = \int_t^T Lu(s, X_{t,x}(s))ds$$
$$+ \int_t^T b(s, X_{t,x}(s))\frac{\partial u}{\partial x}(s, X_{t,x}(s)) \, dW(s).$$

Recalling that $Lu = 0$ and taking expectation of the left- and right-hand sides of the above relation, we get

$$u(t, x) = Eu(T, X_{t,x}(T)) = E\varphi(X_{t,x}(T)).$$

Thus, the solution of the problem (18.26)–(18.27) has the following probabilistic representation

$$u(t, x) = E\varphi(X_{t,x}(T)), \tag{18.29}$$

where $X_{t,x}(s)$, $s \geq t$, is the solution of the SDE (18.28).

---

[8]See the corresponding theoretical results in, e.g., Fridman (1964).

We can generalise (18.29) for the following Cauchy problem for parabolic PDE

$$Lu + c(t,x)u + g(t,x) = 0, \quad t \in [t_0, T), \quad x \in \mathbf{R}, \qquad (18.30)$$

$$u(T,x) = \varphi(x), \quad x \in \mathbf{R}. \qquad (18.31)$$

The solution of this problem has the probabilistic representation:

$$u(t,x) = E\left[\varphi(X_{t,x}(T)) \exp\left\{\int_t^T c(s, X_{t,x}(s))ds\right\} \qquad (18.32)\right.$$
$$\left. + \int_t^T g(s, X_{t,x}(s)) \exp\left\{\int_t^s c(s', X_{t,x}(s'))ds'\right\} ds\right],$$

where $X_{t,x}(s)$, $s \geq t$, is the solution of the SDE (18.28). These probabilistic representations are often called the *Feynman–Kac formula*. It can be also written for multidimensional PDE problems (see, e.g., Wentzell, 1981; Krylov, 2002; Rogers and Williams, 2000; Karatzas and Shreve, 1991; Shreve, 2004).

Remark 18.4. The probabilistic representations of solutions $u(t,x)$ of linear PDEs are the basis for a numerical solution of these PDEs via the Monte Carlo technique (see, e.g., Glasserman, 2003; Milstein and Tretyakov, 2004). The essence of such numerics consists in approximating solutions of the associated SDEs and independent sampling of these approximate SDE solutions, averaging over which gives an approximation of $u(t,x)$.

18.3   The Black–Scholes Equation

Let us start by clarifying the logic we have been following and which we will complete in this section. In Part II we used the discrete binary model (I) together with the no-arbitrage principle and the idea of replication to price European options, which culminated in the discrete Black–Scholes formula (10.7). Then, in Chapter 16 we derived the continuous Black–Scholes formula (16.15) by appropriately choosing the parameters of the binary model (I) and using CLT. Though this already gave us the main result of Part III of this course, we observed that from the modelling perspective it is useful to consider continuous dynamics as well. To this end, we introduced geometric Brownian motion in Chapter 17 and SDEs in Section 18.1 as continuous-time models for financial applications. The remaining link to close the square (see

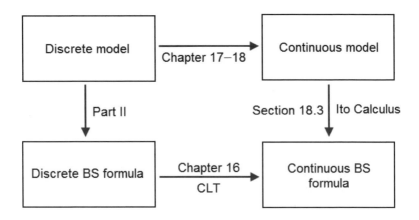

Fig. 18.1   Two derivations of the continuous Black–Scholes formula.

Fig. 18.1) is to derive the continuous Black–Scholes formula starting from the continuous model, which is done in this section using Ito Calculus.

Suppose that, as before, we have two instruments:

$$\text{bank account:} \quad dB = rBdt, \quad B(0) = B_0; \tag{18.33}$$

$$\text{risky-asset:} \quad dS = \mu Sdt + \sigma SdW, \quad S(0) = S_0. \tag{18.34}$$

The model is written under the 'actual' (market) measure.

We see that it is the second term in (18.34) which turns S into a risky asset. If $\sigma = 0$ then S is riskless. Note that the model (18.34) is slightly different to $(17.11)^9$ and (18.9).

Introduce the portfolio $(\phi(t), \psi(t)) = (\phi(t, S(t)), \psi(t, S(t)))$, where $\phi(t)$ is the amount of stock held by the writer at time t and $\psi(t)$ is amount of money held by the writer at time t in the bank account. As usual (see Section 9.4 of Part II), we require that strategies $(\phi(t), \psi(t))$ are $\mathcal{F}_t$-adaptive. The value of the portfolio at time t is

$$V(t) = \phi(t)S(t) + \psi(t) := v(t, S(t)). \tag{18.35}$$

It is convenient to have these two notations for the wealth process.

---

[9]The solution of (18.34) is $S(t) = S_0 e^{(\mu - \frac{\sigma^2}{2})t + \sigma W(t)}$.

As before, we are working with self-financing strategies, i.e. when the portfolio never needs to be topped up with extra cash and neither can it ever afford withdrawals. In the discrete case we had (cf. (9.19)–(9.20)):

$$V(t) = \phi(t)S(t) + \psi(t),$$
$$V(t + \Delta t) = \phi(t)S(t + \Delta t) + \psi(t)B(t + \Delta t)/B(t),$$

and, consequently,

$$V(t + \Delta t) - V(t) = \phi(t)[S(t + \Delta t) - S(t)] + \psi(t)[B(t + \Delta t) - B(t)]/B(t).$$

Taking $\Delta t \to 0$, one can get

$$dV = \phi(t)dS + \psi(t)\frac{dB}{B(t)}. \tag{18.36}$$

Then, in the continuous-time case we say that the *portfolio is self-financing* if it satisfies (18.36).

Substituting (18.33) and (18.34) in (18.36), we obtain

$$dV = \psi(t)rdt + \phi(t)\mu Sdt + \phi(t)\sigma SdW(t). \tag{18.37}$$

Consider a European option with payoff function $f(x)$ and maturity time T. As usual, in the pricing problem we need to find a minimum value of the portfolio

$$y = V(0) = v(0, S(0))$$

such that

$$V(T) = v(T, S(T)) = f(S(T)), \tag{18.38}$$

in which one can see that we are exploiting the idea of replication once again.

Our aim is to get an equation for a smooth function $v(t, x)$ whose solution will give us the option price at time t and the underlier's spot price $S(t) = x$.

Applying the Ito formula to $v(t, S(t))$, we get

$$dv = \left[\frac{\partial v}{\partial t}(t, S) + \frac{\partial v}{\partial x}(t, S)\mu S + \frac{1}{2}\frac{\partial^2 v}{\partial x^2}(t, S)\sigma^2 S^2\right] dt + \frac{\partial v}{\partial x}\sigma SdW(t). \tag{18.39}$$

Comparing (18.37) and (18.39), we see that for any realisation of $S(t)$:

$$\phi(t) = \phi(t, S(t)) = \frac{\partial v}{\partial x}(t, S(t))$$

and

$$\frac{\partial v}{\partial t}(t, S) + \frac{\sigma^2}{2}S^2\frac{\partial^2 v}{\partial x^2}(t, S) = \psi(t, S)r, \quad t < T. \tag{18.40}$$

Due to (18.35), we obtain

$$\psi = V - \phi S = v(t, S) - \frac{\partial v}{\partial x}(t, S)S. \tag{18.41}$$

Substituting (18.41) in (18.40), we get

$$\frac{\partial v}{\partial t}(t, S) + \frac{\sigma^2}{2}S^2\frac{\partial^2 v}{\partial x^2}(t, S) = rv(t, S) - rS\frac{\partial v}{\partial x}(t, S),$$

which should hold for any realisation of S and at every $t < T$. Consequently, it should be true for any $x > 0$. Hence we arrive at the following (deterministic) equation for $v(t, x)$:

$$\frac{\partial v}{\partial t} + \frac{\sigma^2}{2}x^2\frac{\partial^2 v}{\partial x^2} + rx\frac{\partial v}{\partial x} - rv = 0, \quad t < T, \quad x > 0, \tag{18.42}$$

with the terminal condition

$$v(T, x) = f(x), \quad x > 0. \tag{18.43}$$

We again eliminated randomness in option pricing.

The derived (18.42)–(18.43) is the Cauchy problem for linear parabolic PDE, which is the famous *Black–Scholes equation*. Its solution gives us the price of a European option with payoff $f(x)$ and maturity at time T.

For clarity, let us summarise the assumptions under which this equation has been derived:

- No arbitrage opportunity.
- It is possible to borrow and lend money in any amounts at a known constant risk-free interest rate r.
- We can buy and sell the stock in unlimited amounts for the existing spot price.
- There is no charge for holding a positive or negative amount of stock.
- Trading is continuous.
- Stock price follows geometric Brownian motion (constant drift and volatility).
- Stock does not pay dividends.
- No taxes and no transaction costs.

'The Black–Scholes assumptions are famously poor. Nevertheless my practical experience of seeking arbitrage opportunities, and my research on costs, hedging errors, volatility modelling and fat tails, for example, suggest that you won't go far wrong using basic Black–Scholes, perhaps with the smallest of adjustments, either for pricing new instruments or for exploiting mispriced options.' (Wilmott, 2008)

Remark 18.5. Here, we presented the derivation of the Black–Scholes equation following the paper by Milstein and Schoenmakers (2002). One can find its other derivations, e.g., in Björk (2004); Shreve (2004) and Shiryaev (1999).

Remark 18.6. The trend μ from the asset model (18.34) does not appear in the Black–Scholes equation (18.42)–(18.43) (see also Remark 16.2).

If we solve (18.42)–(18.43), we will get the function $v(t,x)$ and, in particular $v(0,x)$, which gives the fair price for a European option at time $t = 0$ by substituting $x = S(0)$. The Cauchy problem (18.42)–(18.43) can be solved analytically. For instance, in the case of the European call with $f(x) = (x - K)_+$ we have[10]

$$v(t,x) = x\Phi\left(\frac{\ln\frac{x}{K} + (r + \frac{\sigma^2}{2})(T - t)}{\sigma\sqrt{T - t}}\right) \tag{18.44}$$

$$- Ke^{-r(T-t)}\Phi\left(\frac{\ln\frac{x}{K} + (r - \frac{\sigma^2}{2})(T - t)}{\sigma\sqrt{T - t}}\right).$$

This Black–Scholes formula (18.44) for the price of a European call coincides with (16.16) from Section 16.2 which was obtained as a limit of the discrete-time model.

Exercise 18.3. Suppose the interest rate r is small and an ATM call option is close to its maturity (i.e. $T - t$ is also small). Find an approximation (in analytic form) of the option's price.

Example 18.5. Write a probabilistic representation for the Black–Scholes problem (18.42)–(18.43).

Solution. Introduce the SDE

$$dS = rSdt + \sigma SdW. \tag{18.45}$$

According to (18.32), the probabilistic representation for $v(t,x)$ has the form

$$v(t,x) = E\left[f(S_{t,x}(T))\exp\left\{-\int_t^T rds\right\}\right] \tag{18.46}$$

$$= e^{-r(T-t)}Ef(S_{t,x}(T)),$$

---

[10]To check that (18.44) is the solution to (18.42)–(18.43), you can substitute it back into (18.42)–(18.43) and confirm that the $v(t,x)$ satisfies the Cauchy problem. See also Problem 22 in Chapter 19 for practicing how this Cauchy problem can be solved. Let us note that it is beneficial to have a good background in PDEs for further study of Stochastic Analysis and Financial Mathematics, where it is useful to know both analytical and numerical PDE techniques. See for background on PDEs, e.g., Evans (2010); Fridman (1964).

where $S_{t,x}(T)$, $s \geq t$, is a solution of (18.45) with the initial condition $S(t) = x$ and averaging is over realisations of the $S_{t,x}(T)$. We again pay attention to the fact that the equation (18.45) (in comparison with the model for the stock price (18.34) written under the market measure) does not depend on μ; it is written under the risk-neutral measure. In this respect it is useful to re-read Remark 17.4.

Remark 18.7. We see that $S_{t,x}(s)$ from (18.45) is a geometric Brownian motion under the risk-neutral measure and

$$S_{t,x}(T) = xe^{(r-\frac{\sigma^2}{2})(T-t)+\sigma(W(T)-W(t))}$$

(compare it with (17.12)). In particular, $S_{0,S_0}(T) = S_0 e^{(r-\frac{\sigma^2}{2})T+\sigma W(T)}$ and the price of the European option at time $t = 0$ can be written as

$$v(0, S_0) = e^{-rT}Ef(S_{0,S_0}(T)) = e^{-rT}E[f(S_0 e^\theta)], \qquad (18.47)$$

where $\theta \sim \mathcal{N}((r - \frac{\sigma^2}{2})T, \sigma^2 T)$. In Section 16.2 we obtained the same result (see (16.15)) as the limit of the discrete-time model (see also Remark 16.1).

Remark 18.8. (*Market price of risk*). Recall that in (18.33) r is the growth rate of money on the riskless bank account and in (18.34) the parameter μ is the growth rate of the considered asset while σ is its risk measure. Then the ratio

$$\gamma = \frac{\mu - r}{\sigma}$$

is the excess rate (i.e. above the risk-free rate r) of return per unit of risk and it is called *the market price of risk*. If we would simultaneously consider several risky assets, the market price of risk is the same for all of these assets which is one of the important consequences of risk-neutral pricing (Baxter and Rennie, 1996; Björk, 2004).

Let us now construct the *hedging strategy*. As we have already found, the delta (i.e. the ratio of change in the price of the option and change in the price of underlying stock, see Section 10.2) in the continuous-time case is equal to

$$\Delta(t) := \phi(t) = \phi(t, S(t)) = \frac{\partial v}{\partial x}(t, S(t)), \qquad (18.48)$$

where $v(t, x)$ is the solution of (18.42)–(18.43). As in the discrete case, the writer charges the holder the premium $x = v(0, S(0))$ at time $t = 0$. Using this money, he forms the portfolio $(\phi(0), \psi(0))$ with $\phi(0)$ from (18.48) and $\psi(0)$ such that (see (18.41))

$$\psi(0) = v(0, S(0)) - \Delta(0)S(0).$$

Then, at each time $t \leq T$, the writer should adjust his portfolio $(\phi(t), \psi(t))$ using the wealth $v(t, S(t))$ so that $\phi(t)$ is from (18.48) and

$$\psi(t) = v(t, S(t)) - \Delta(t)S(t). \tag{18.49}$$

This strategy (called *delta hedging*[11]) results in the writer having the portfolio at time T whose value matches his obligation $f(S(T))$. It is not difficult to see that this strategy is a limit of the hedging strategy we had in the discrete case (see Section 10.2).

18.4   Sensitivities or Greeks

Previously we introduced and used for hedging, one of the price sensitivities (*Greeks*), delta. In particular, in our Black–Scholes case we can find deltas explicitly for simple derivatives.

Example 18.6. In the case of the European call the delta is equal to

$$\Delta(t) = \frac{\partial v}{\partial x}(t, S(t)) = \Phi\left(\frac{\ln \frac{S(t)}{K} + (r + \frac{\sigma^2}{2})(T - t)}{\sigma\sqrt{T - t}}\right). \tag{18.50}$$

Exercise 18.4. Show that (18.50) is true.

Exercise 18.5. Plot Δ from (18.50) as a function of the underlier's price for different times t to the maturity T.

Exercise 18.6. Using the previous exercise or otherwise, explain what happens to delta of an ATM call as a function of S when time t approaches maturity T. What does your observation imply concerning hedging of such an option?

Practitioners also use many other Greeks including the gamma Γ, theta Θ, rho ρ and vega[12] $\mathcal{V}$:

$$\Gamma = \frac{\partial^2 v}{\partial x^2}, \quad \Theta = \frac{\partial v}{\partial t}, \quad \rho = \frac{\partial v}{\partial r}, \quad \mathcal{V} = \frac{\partial v}{\partial \sigma},$$

which are the sensitivities of the derivative price (or more generally, of a portfolio) to the corresponding parameters of the market. For instance, we see that ρ indicates the change in option value with the change in the interest

---

[11]It is an example of dynamic hedging, of which another example – gamma hedging – is considered in the next section.

[12]Although there is no 'vega' in the Greek alphabet.

rate, $\mathcal{V}$ gives the change in option value with the change in the volatility, Θ reflects the time-decay of the option price, etc.

Example 18.7. The delta $\Delta_S(t)$ of the underlying stock is equal to 1 and its gamma $\Gamma_S(t) = 0$.

Example 18.8. Observe that the Black–Scholes equation (18.42) can be rewritten using Greeks in the following form:

$$\Theta(t) + \frac{\sigma^2}{2}x^2\Gamma(t) + rx\Delta(t) - rv = 0. \tag{18.51}$$

Greeks are used in practice (see, e.g., Björk, 2004) for monitoring the sensitivity of a portfolio (consisting of stock and derivatives) with respect to: (i) the change of price of the underliers and (ii) changes in parameters of the model used for pricing. The first is needed for monitoring how the portfolio value varies with a change in the price of the underlying stock and is important for risk management of the portfolio (for hedging). The second gives sensitivity with respect to mis-specifications of the model parameters.

Definition 18.1. *If a certain Greek of a portfolio is equal to zero, then this portfolio is called **neutral** with respect to this Greek.*

For instance, a portfolio with zero delta is called *delta neutral*.

Example 18.9. Let $V(t)$ be the value of a portfolio with a single underlying stock (i.e. the portfolio can contain the stock itself and derivatives written on it). Compose the new portfolio with value $U(t)$, which consists of the original portfolio and z amount of a derivative with price $v(t)$, written on the same underlier, i.e.

$$U(t) = V(t) + zv(t). \tag{18.52}$$

The new portfolio is delta neutral (please check) if $z = -\Delta_V/\Delta_v$ (here Δ_V is the delta of the original portfolio and Δ_v is the delta of the derivative).

Example 18.10. Suppose our portfolio consists of a single derivative we sold, for which the current price is $v(t, x)$, then $V(t) = -v(t, x)$ is the current value of our portfolio. In the above example we add the underlier instead of the derivative to the original portfolio to achieve a delta neutral portfolio and the new portfolio has the value $U(t) = -v(t, x) + zx$, which is delta neutral if $z = \Delta_v = \partial v(t, x)/\partial x$ – the amount of stock we need to hedge the derivative, which is the result we already know from the previous section.

In the terminology used in practice, the requirement in Definition 18.1 for a Greek to be exactly equal to zero is relaxed. That is, one says that a portfolio is neutral with respect to a sensitivity to a parameter if the portfolio is not affected by small changes in this parameter, i.e. if the Greek is small in a practitioner's judgement, she says that the portfolio is neutral with respect to it.

From the previous section, we, in particular, learned that in the Black–Scholes world (and, in fact, in all standard continuous-time models) delta hedging should be done continuously. Though this continuous rebalancing of the portfolio ensures (at least theoretically) the perfect hedge, it is impractical to do hedging continuously. Amongst reasons for this is that in reality there are transaction costs to be paid for any rebalancing. A trader prefers to avoid extra costs and hence she does not want to carry out hedging too often. The need for rebalancing comes from the varying of delta when the price of the underlier changes. If delta does not significantly change, our portfolio stays delta neutral. If delta changes, we have to rebalance the portfolio. Gamma is measuring this sensitivity of delta. If gamma is small, then delta stays almost constant (and hence we can avoid buying/selling stock for the rebalancing of the portfolio). Thus, it is of interest to attempt to construct a portfolio which is neutral both with respect to delta and gamma.

Example 18.11. (*Gamma hedging*). Let us construct a portfolio which is neutral both with respect to delta and gamma. As in Example 18.9, suppose $V(t)$ is the value of a portfolio with a single underlying stock. Compose a new portfolio with value $U(t)$ consisting of the original portfolio and z amount of a derivative with price $v(t)$ and y amount of a derivative with price $u(t)$ both written on the same underlier, i.e.

$$U(t) = V(t) + zv(t) + yu(t). \qquad (18.53)$$

(As these two derivatives, we can take, e.g., two puts with different strikes and the same maturity). Due to our objective, we require that both first and second derivatives of U with respect to the underlier's price are equal to zero. Then (please check) the following values of z and y make the new portfolio neutral with respect to delta and gamma:

$$y = \frac{\Delta_V \Gamma_v - \Delta_v \Gamma_V}{\Delta_v \Gamma_u - \Delta_u \Gamma_v}, \quad z = \frac{\Delta_V \Gamma_u - \Delta_u \Gamma_V}{\Delta_u \Gamma_v - \Delta_v \Gamma_u}, \qquad (18.54)$$

where the notation is analogous to the one used in Example 18.9.

We can simplify (18.54) by using stock in (18.53) instead of the second option. Then, due to Example 18.7, $\Delta_u = 1$ and $\Gamma_u = 0$. This allows us to have a two-stage hedging procedure: first choose $z = -\Gamma_V/\Gamma_v$ to make the portfolio consisting of the original portfolio with value V and an option with price v to be gamma neutral; second add the amount $y = \Delta_v \Gamma_V/\Gamma_v - \Delta_V$ of the stock. The second stage makes the final portfolio delta neutral and at the same time continues to keep it gamma neutral as well.

Example 18.12. (*With numbers*). An investor has a portfolio in which delta is equal to -3 and gamma is equal to 1. There is an option traded on the market where delta is equal to 0.5 and gamma is equal to 0.1. How can the investor construct a hedging portfolio which is delta and gamma neutral? *Answer for verification*: the investor holding the original portfolio should short sell ten options and buy eight units of stock.

Remark 18.9. Note that if we live in the Black–Scholes world and we hold a portfolio which satisfies the Black–Scholes equation (18.42) and this portfolio is neutral with respect to delta and gamma then it follows from (18.51) that this portfolio has the riskless rate of return equal to the interest rate r.

For further reading on Greeks, see Bingham and Kiesel (2004); Björk (2004); Kolb (2003) and Musiela and Rutkowski (2005).

18.5   Discussion on the Use of Modelling in Financial Engineering

There are two main areas where modelling is used for pricing financial derivatives:

1. Finding current prices and hedges of derivatives which are not traded on exchanges or where trading is sparse.
2. Finding future prices of derivatives.

The practical need for the first area comes from a regulatory requirement for marking-to-market a financial firm's portfolio of derivatives on a regular basis and from risk management of the portfolio. Because the majority of derivatives in a firm's portfolios are usually OTC, one has to run a model to estimate their current prices and hedges (the latter are important for risk management as we know from the course). The model for pricing a particular derivative is calibrated (first of all, the corresponding volatility surface) to the current state of the market via observing liquid derivatives tradable on an exchange which are written on the same underlier as the derivative with unobservable price. This is not always possible, e.g., a company with which a bank entered into a swap might not be listed on an exchange or its bonds or shares have low liquidity. Then, alternatively, other assets (often derivatives on stock market indices such as FTSE 100, S&P 500, Nikkei 225, etc.), which are assumed (with support of statistical analysis of historical data and backtesting) to be strongly correlated with the product in question, are used for calibration. Other financial information about the underlier can also be exploited in calibration (e.g., information on dividend payments and corporate bond prices of a company to which the underlier is related, credit ratings of a company and financial products, etc.). Further, where market discovery of prices is limited, *consensus market prices*[13] are used for calibration.

The second area, finding future prices of financial products, is important for risk assessment of a portfolio and of a firm. A firm has to regularly evaluate risk measures related to the probability of its loss or default over some time horizon. Examples of such measures include Value at Risk[14] (VaR) and its variants (Jorion, 2006), potential future exposure (PFE), expected positive exposure (EPE) or expected shortfall, also called expected tail loss, within a given future time period (e.g., 12 months from now) (Cesari *et al.*, 2010). This assessment determines, in particular, a capital requirement on the firm and it is a part of, e.g., the Basel Accord (see the Basel Committee on Banking Supervision website http://www.bis.org/bcbs/). It is also closely related to computing a price of hedging of counterparty risk, which is generally referred to as credit valuation adjustment (CVA) (Cesari *et al.*, 2010). To find future prices even of derivatives tradable on an exchange, one has to use models. Calibration of the models in this case is done similarly to as was discussed in the previous paragraph but the task of evaluating the described risk measures is much more complex (especially computationally) than the task from the first area discussed above. Indeed, to solve this problem, we first need to run a model for the underlier $S(t)$ under the 'real world' probability measure until the future time t_* at which a risk measure or CVA

---

[13]They are provided by the Totem service which works in the following way. Major market players in OTC derivatives report prices of exotic products to Totem every month. In return, Totem supplies the firms with average prices for such products together with a spread after dropping prices that are considered to be outliers (this methodology is, to some extent, similar to setting LIBOR by BBA).

[14]Let η be a random variable modelling the mark-to-market loss of a portfolio over some time period and α be a given confidence level. Then VaR is defined as the α-quantile for $\eta : VaR_\alpha(\eta) = \inf_{x \in \mathbf{R}} \{ P(\eta > x) \leq 1 - \alpha \}$, where P is a market probability measure.

should be estimated. Then, using the found sample of the underlier's prices $S(t_*)$ at t_*, we need to evaluate prices of derivatives $v(t_*, S(t_*))$[15] at the set of $(t_*, S(t_*))$, which requires running the model further, from t_* until a maturity T of the derivative, under an EMM. As a result, a sample distribution of $v(t_*, S(t_*))$ will be obtained (see further details in Cesari *et al.*, 2010).

To conclude, even within this introductory course we have seen that the trading of financial derivatives is very complex and the use of models for better decision making, risk managing and also training of traders and quants cannot be underestimated.

$$\Diamond \ \Diamond \qquad \mathcal{THE\ END} \qquad \Diamond \ \Diamond$$

A fairy tale does not finish at its end.
Remember, we asked you at the beginning,
'What would remain from a fairy tale,
After it has been told?'[16]

---

[15]The notation here has the same meaning as in (18.46).
[16]Translated from the final song of Visotsky's audioplay; see Footnote 1 on p. 5.

Chapter 19

Problems for Part III

1. (*Probability revision*). Prove that if $\xi \sim \mathcal{N}(m, \gamma^2)$ and a, $b \in \mathbf{R}$ then $a\xi + b \sim \mathcal{N}(am + b, a^2\gamma^2)$.

2. (*Probability revision*). Let ξ be a random variable, which is normally distributed with mean m and standard deviation σ. Show that
$$E \exp(a\xi) = \exp\left(am + \frac{\sigma^2 a^2}{2}\right), \quad a, \sigma \in \mathbf{R}.$$

3. (*Probability revision*). Let a random variable ξ be such that $\log \xi = \eta$, where η is a Gaussian random variable with mean μ and variance σ^2. Deduce the distribution function of ξ and its density (the density of log-normal distribution).

4. Consider the discrete-time process S_n defined as
$$S_n = S_{n-1} + \mu \frac{S_{n-1} + S_n}{2}\Delta t + \sigma S_{n-1}\sqrt{\Delta t}\xi_n, \quad n = 1, \ldots, N,$$
where μ, $S_0 > 0$, $\sigma > 0$ are constants, ξ_n are i.i.d. random variables with the law $P(\xi_n = \pm 1) = 1/2$, $T = \Delta t N$, and Δt is sufficiently small. Find the limit distribution of S_N (under the given measure P) when $\Delta t \to 0$.

5. Consider the discrete-time process S_n defined as
$$S_n = S_{n-1} + \sigma S_n\sqrt{\Delta t}\xi_n, \quad n = 1, \ldots, N,$$
where $S_0 > 0$, $\sigma > 0$ are constants, ξ_n are i.i.d. random variables with the law
$$P(\xi_n = 0) = \frac{2}{3}, \quad P\left(\xi_n = \pm\sqrt{3}\right) = \frac{1}{6}.$$
Assuming $T = \Delta t N$, find the limit distribution of S_N when $\Delta t \to 0$.

6. (*The Black–Scholes formula for pricing European call*). Consider the expectation (see Example 16.4):
$$C_0 = E\left[\left(S_0 \exp\left(-\frac{\sigma^2}{2}T + \sigma\sqrt{T}\tilde{\theta}\right) - Ke^{-rT}\right)_+\right],$$

where $\tilde{\theta}$ is a Gaussian random variable with zero mean and unit variance while K is a strike price, T is the time to the maturity, r is the interest rate, σ is the volatility and S_0 is the spot price of the underlying. By computing the expectation, prove (16.16).

7. What is the probability that at a fixed time $t > 0$ the standard Wiener process $W(t)$ has a positive value?

8. Let $W(t)$ be a standard Wiener process. Compute the mean and variance of $4W(4) + W(1)$.

9. Let $\{\mathcal{F}_t\}_{t\geq 0}$ be a natural filtration for the standard Wiener process $W(t)$. Which of the following stochastic processes are adapted to the filtration $\{\mathcal{F}_t\}_{t\geq 0}$ and which are not?
   (a) $Y(t) := \max_{t \leq s \leq T} W(s)$;
   (b) $Z(t) := \min_{0 \leq s \leq t} W(s)$.

10. Let $W(t)$ be a standard Wiener process. Compute the covariances
    (a) $Cov(3W(3) + W(2), W(1))$;
    (b) $Cov(W(s), W(t))$;
    (c) $Cov(W(t), W^2(t))$.

11. Let $W(t)$ be a standard Wiener process. Fix some $s \geq 0$. Show that the process $\tilde{W}(t) := W(t + s) - W(s)$, $t \geq 0$, is also a standard Wiener process.

12. Let ξ be a random variable. The function $\varphi(\lambda) := E\left[\exp(\lambda \xi)\right]$ is called the *moment generating function* of ξ.
    (i) Find the moment generating function for a Gaussian random variable with zero mean and variance b^2.
    (ii) One can prove (Shiryaev, 1996; Williams, 2001) that the rth moment of ξ is equal to the rth derivative of the moment generating function $\varphi(\lambda)$ at $\lambda = 0$. Using this fact and your answer on (i), show that

$$E\left[W^4(t)\right] = 3t^2,$$

where $W(t)$ be a standard Wiener process.

13. Let $W(t)$ be a standard Wiener process. Compute

$$E\int_0^t W^2(s)dW(s) \quad \text{and} \quad E\int_0^t W^2(s)ds.$$

14. Using an appropriate property of Ito integrals, compute

$$E\left[\int_0^t W_1(s)dW_2(s)\right]^2,$$

where $W_1(t)$ and $W_2(t)$ are independent standard Wiener processes.

15. Let $W_1(t)$ and $W_2(t)$ be independent standard Wiener processes. Using the definition of Ito integral, show that

$$\int_0^t W_1(s)dW_2(s) = W_1(t)W_2(t) - \int_0^t W_2(s)dW_1(s).$$

16. Using the Ito formula, show that

$$\int_0^t sdW(s) = tW(t) - \int_0^t W(s)ds$$

(here $W(s)$ is a standard Wiener process).

17. Let $g_n(x) = x^n$, where n is a positive integer, and let $W(t)$ be a standard Wiener process. Using the Ito formula, derive the equation for $\xi^{(n)}(t) := g_n(W(t))$ and compute $EW^{2n}(t)$.

18. Let $S(t)$ satisfy the SDE:

$$dS = \mu(t)Sdt + \sigma(t)SdW(t), \quad S(0) = S_0,$$

with some drift $\mu(t)$ and volatility $\sigma(t)$ which are $\mathcal{F}_t$-adapted, continuous functions of t and $E\sigma^2(t)S^2$ is bounded. Introduce the process $X(t) = \ln S(t)$. Using the Ito formula, find an SDE for $X(t)$.

19. Exploiting the Ito formula, write an SDE for the process $X(t) = \cos(W(t))$, where $W(t)$ is a standard Wiener process, and compute $EX(2)$.

20. (*Ornstein–Uhlenbeck process*). Consider the linear SDE with additive noise

$$dX = -\alpha Xdt + \sigma dW(t), \quad X(0) = x.$$

Using the Ito formula, show that this SDE has the solution:

$$X(t) = xe^{-\alpha t} + \sigma \int_0^t e^{\alpha(s-t)}dW(s).$$

21. The Cox–Ingersoll–Ross model for the short rate $r(t)$ can be written in the form of the following Ito SDE:

$$dr = a(b - cr)dt + \sigma\sqrt{r}dW(t), \quad r(0) = r_0.$$

Compute the expectation $Er(t)$ and variance $Var(r(t))$. Also find $\lim_{t\to\infty} Er(t)$.

22. Using PDE techniques (without using any probabilistic argument), find the solution of the Cauchy problem (18.42)–(18.43) with $f(x) = (x-K)_+$ and thus confirm (18.44).

23. Consider the Cauchy problem for a PDE

$$\frac{\partial u}{\partial t} + \frac{1}{2}\frac{\partial^2 u}{\partial x^2} - cu = 0, \quad t \in [0,1), \; x \in \mathbf{R},$$

$$u(1,x) = x, \quad x \in \mathbf{R},$$

where c is some constant.

(a) Write a probabilistic representation of the solution $u(t,x)$.

(b) Using your answer to part (a) of this question, find the function $u(t,x)$.

24. In the Black–Scholes world the price of a European plain vanilla call on a stock that does not pay dividends is equal to $C_0 = v(0, S_0)$ with $v(t,x)$ from (18.44).

(a) Explain what S_0, r, σ, K and T in this formula are.

(b) Compute the vega of this option.

(c) Confirm that the vega you have computed in the part (b) of this question is always positive.

(d) Give a financial explanation for (c).

(e) Plot the vega of this option as a function of the underlier's price and different maturities T.

(f) Establish what happens to vega when T goes to zero (i.e. when an option is close to its maturity). Give a financial explanation for your answer.

25. In the Black–Scholes world the price of a European plain vanilla put on a stock that does not pay dividends is equal to

$$D_0 = Ke^{-rT}\Phi\left(\frac{\ln(K/S_0) - (r - \sigma^2/2)T}{\sigma\sqrt{T}}\right)$$

$$- S_0\Phi\left(\frac{\ln(K/S_0) - (r + \sigma^2/2)T}{\sigma\sqrt{T}}\right),$$

where $\Phi(x)$ is the normal distribution function.

(a) Compute the delta of this option.

(b) Is it possible to say whether the found delta is always positive or negative?

(c) Give a financial explanation for the observation made in the part (b) of this question.

(d) Plot Δ of this option as a function of the underlier's price for different maturities and strikes. Make observations about delta behaviour with increase of strike and decrease of time to maturity.

26. Show that gamma of the European call coincides with gamma of the European put with the same strike, maturity and written on the same underlying stock.

27. Consider an ATM European call or put option. What happens to its gamma when the option is close to its maturity? Then discuss why it is difficult to hedge ATM options close to their maturity.

28. (*Vega hedging*). Consider a portfolio with a single underlying stock (i.e. the portfolio that can contain the stock itself and derivatives written on it). Form a portfolio, which consists of the original portfolio and possibly some derivatives written on the same stock and/or the underlying stock itself, so that this new portfolio is neutral with respect to both delta and vega.

Chapter 20

Further Study

The field of Financial Mathematics and Financial Engineering is very large and rapidly developing. This chapter is about what has not found a place in this book, mainly due to the time restriction of a one-semester course but also due to the very low level of the prerequisites.

The course has roughly been about two areas: Finance and Stochastics. It is natural to say separately what has not been considered in each of them.

We start with *Stochastics*. First, the coverage of Probability was rather narrow. For further study Shiryaev (1996) and Williams (2001) can be used. Second, the introduction to stochastic processes and Ito Calculus was (deliberately) done on a naïve level. Here, knowledge can be enhanced using the introductory textbooks Klebaner (2005) and Mao (2007) and more comprehensive ones such as Wentzell (1981); Karatzas and Shreve (1991); Krylov (2002) and Rogers and Williams (2000). Third, Poisson and Levy processes were not covered at all, though they play an important role in financial engineering modelling. As we know, in the Black–Scholes world, the share price is continuous in time. However, unexpected drastic events like a significant political change, natural disaster, a report of unexpected economic data, departure of a key leader from a company or any other sudden 'bad' or 'good' news often lead to a jump in prices. Then in financial modelling, especially on a shorter time scale, a Poisson process has to be added to Brownian motion driven dynamics to adequately represent price behaviour (see, e.g., Bingham and Kiesel, 2004; Lamberton and Lapeyre, 2007; Shreve, 2004). The Wiener process and Poisson process belong to a wide class of stochastic processes with independent, stationary increments, which are called Levy processes. Stochastic differential equations driven by Levy processes (Applebaum, 2009) are amongst the popular models in financial engineering[1].

---

[1] See Carr *et al.* (2002) and also Cont and Tankov (2003) and references therein.

There are other mathematical and related quantitative areas for possible further study. We saw that the theory of PDEs plays an important role in Financial Mathematics, which can be studied using Fridman (1964) and Evans (2010). Statistics and Data Mining techniques are used for estimating parameters and financial time series analysis. As a starting point on these topics, see Chapters 8 and 9 in Dokuchaev (2007). A more comprehensive coverage requires a separate course (see, e.g., Ruppert, 2011).

Finally, numerical aspects were not a part of the course but they, together with programming skills and an understanding of complex computer systems, are key for the successful career of a quantitative risk analyst (quant) in the financial industry. As an initial textbook on Numerical Analysis covering a wide range of topics, Cheney and Kincaid (2012) can be used. Numerical methods for stochastic differential equations are considered in Glasserman (2003); Milstein and Tretyakov (2004) and Cont and Tankov (2003). For object-oriented programming and C++ tailored towards Finance, see, e.g., Duffy (2006); Joshi (2008) and Schlogl (2013).

Now, *Finance.* First, the coverage of fixed-income products was rather limited. To progress further in this area, one can use introductory/intermediate level textbooks such as Filipovic (2009); Björk (2004) and Musiela and Rutkowski (2005), while Brigo and Mercurio (2006) and Andersen and Piterbarg (2010) are comprehensive books on the subject, including the practical aspects of fixed-income markets.

Second, the course was dealing mainly with derivatives on commodities and equity but we considered the simplest models only. We discussed that one of the drawbacks of the geometric Brownian motion is the assumption of constant volatility and this model cannot capture, e.g., the volatility smile phenomenon. A generalisation of the geometric Brownian motion includes a variety of stochastic volatility models (Bingham and Kiesel, 2004; Musiela and Rutkowski, 2005; Shreve, 2004) and also jump-diffusion models (Bingham and Kiesel, 2004; Lamberton and Lapeyre, 2007; Shreve, 2004) as we already mentioned earlier in this chapter.

Thirdly, we did not touch on such an important area in modern finance as credit risk and credit derivatives. The starting point could be chapters on these topics in Bingham and Kiesel (2004); Brigo and Mercurio (2006) and Lamberton and Lapeyre (2007). Deeper coverage can be found in Bielecki and Rutkowski (2010); Schmid (2004) and Lipton and Rennie (2008). The importance of the area of counterparty risk was especially highlighted following the financial crisis of 2007–2008. It includes such elements as measuring counterparty exposure, estimating capital requirements and hedging counterparty risk, which are crucial for the stability and security of a bank (Cesari *et al.*, 2010).

Further, we did not consider portfolio management and, in particular, the two popular theories: Markowitz portfolio optimisation and CAPM (capital asset pricing model). In managing a portfolio, one is usually interested in either maximising return under some restriction on the riskiness of the portfolio or minimising the risk of the portfolio. This is achieved via spreading the investment among several securities which is called diversification. Though intuitively we understand what risk is, for making decisions we need a measure of risk (Delbaen, 2003; Holton, 2004). Markowitz's theory uses expected return (i.e. expectation of a future value of the portfolio) as the measure for return and variance as the measure of risk. One can start studying the modern portfolio theory using Chapter 2 of Shiryaev (1999), Chapter 5 of Capinski and Zastawniak (2003), Barucci (2003) and Elton *et al.* (2009). Modern aspects of risk measures are available, e.g., in Föllmer and Schied (2004).

Other areas of Financial Mathematics and Engineering include their links with insurance (actuarial) mathematics (Melnikov, 2004) and also include the regulatory framework for derivative trading and banking in general, of which some elements are covered, e.g., in Hull (2003), Kolb (2003) and Choudhry (2012).

Chapter 21

Appendix: Solutions

The Appendix contains solutions and/or answers to most of the problems given at the end of Parts I (Chapter 6), II (Chapter 13) and III (Chapter 19) of the book. In the preparation of these problems, Hull (2003); Shreve (2003); Stefanica (2011) and Zhou (2008) were partially used. Further problems can be found, e.g., in Etheridge (2002); Björk (2004); Hull (2003, 2006); Dokuchaev (2007); Shreve (2003); Stefanica (2011); Zhou (2008) and Filipovic (2009).

21.1 Solutions to Problems for Part I

1. We need to find the level monthly payment P so that

$$200000e^{25 \times 0.05} - P \sum_{j=0}^{25 \times 12 - 1} e^{j \times 0.05/12} = 0.$$

Noting that $\sum_{j=0}^{25 \times 12 - 1} e^{j \times 0.05/12} = \dfrac{e^{25 \times 0.05} - 1}{e^{0.05/12} - 1}$, we have

$$P = \frac{200000e^{25 \times 0.05}}{e^{25 \times 0.05} - 1}(e^{0.05/12} - 1) \doteq 1170.40 \ (\pounds).$$

2. We have: the principal $B_0 = \pounds 100{,}000$, the final balance $B_5 = \pounds 111{,}529$.
(a) Due to an interest rate with annual compounding,

$$B_5 = B_0 (1 + r_1)^5.$$

Then

$$r_1 = \left(\frac{111529}{100000}\right)^{1/5} - 1 \doteq 2.206 \times 10^{-2} \ (2.206\%).$$

(b) Due to an interest rate with monthly compounding,

$$B_5 = B_0 \left(1 + \frac{r_{12}}{12}\right)^{5 \times 12}.$$

Then

$$r_{12} = 12 \times \left[\left(\frac{111529}{100000}\right)^{1/60} - 1\right] \doteq 2.184 \times 10^{-2} \quad (2.184\%).$$

(c) Due to an interest rate with continuous compounding,

$$B_5 = B_0 e^{5r}.$$

Then

$$r = \frac{1}{5} \ln\left(\frac{111529}{100000}\right) \doteq 2.182 \times 10^{-2} \quad (2.182\%).$$

3. We have the current balance $B_5 = £50{,}000$.
(i) Then $X = B_0 = 50000 e^{-0.08 \times 5} \doteq 33516$ (£).
(ii) We have $B_{10} = 50000 e^{0.08 \times 5} \doteq 74591$ (£).
4. 3.5% interest rate with quarterly compounding means that the balance after one year is

$$B_1 = B_0 (1 + \frac{0.035}{4})^4,$$

where B_0 is a principal.
(a) The rate r_{365} with annual compounding means that

$$B_1 = B_0 (1 + \frac{r_{365}}{365})^{365}.$$

Then

$$r_{365} = 365 \left[(1 + \frac{0.035}{4})^{4/365} - 1\right] \doteq 0.03485 \quad (3.485\%).$$

(b) The rate r_{12} with monthly compounding means that

$$B_1 = B_0 (1 + \frac{r_{12}}{12})^{12}.$$

Then

$$r_{12} = 12 \left[(1 + \frac{0.035}{4})^{1/3} - 1\right] \doteq 0.03490 \quad (3.490\%).$$

(c) The rate r with continuous compounding means that $B_1 = B_0 e^r$. Then, we have to equalise

$$B_0 (1 + \frac{0.035}{4})^4 = B_0 e^r.$$

Hence

$$r = 4\ln(1 + \frac{0.035}{4}) \doteq 0.03485 \ (3.485\%).$$

5. (a) The forward price is

$$K_0 = £145.05 \times 100 \times e^{0.005} \doteq £14577.7.$$

(b) The present forward price (six months later) is

$$K_1 = £151.96 \times 100 \times e^{0.0125/2} \doteq £15291.3.$$

The present value of the original short forward contract

$$e^{-0.0125/2}(S_T - K_0 - [S_T - K_1]) = e^{-0.0125/2}(K_1 - K_0)$$
$$\doteq e^{-0.0125/2} \times (15291.3 - 14577.7) \doteq 709.1 \ (£).$$

(c) The forward price is too low. This creates, e.g., the following arbitrage opportunity.

- At time $t = 0$

  - enter into the long forward on 100 tonnes of feed wheat with the forward price $K = £14,500$;
  - borrow 100 tonnes of feed wheat and sell for £14,505;
  - put £14,505 on the bank account with the interest rate 0.5% p.a. with continuous compounding.

  The value of this portfolio is zero.
- At time $T = 1$

  - take $£14,505e^{0.005} = £14,577.7$ from the bank account;
  - using the long forward buy 100 tonnes of feed wheat for $K = £14,500$ using the money you got from the bank.

  Your riskless profit is $14577.7 - 14500 = 77.7 \ (£)$.

(d) Under these new conditions, the forward price is

$$K_0 = (145.05 + 0.25) \times 100 \times e^{0.005-0.0025} \doteq 14566.4.$$

6. E.g., the following solution can be accepted (a correct answer with an argument without strategies or with a different strategy can also be accepted). To discover the fair forward price, consider the following strategies.

- At time $t = 0$

  - borrow $\$100 \times e^{-0.04} \doteq \$96.079$;

– using this money, buy $e^{-0.04} \doteq 0.96079$ of the share.
During $0 \leq t \leq 1$ invest the yield on shares in the shares. Then

• At time $t = T = 1$ you

– have one share;
– should return to the bank $K_* = 100 \times e^{-0.04} \times e^{0.02} \doteq 98.020$ ($).

Hence, the forward price $K \leq K_*$ (by this scheme you can get the share by paying K_* at $t = 1$, and then if $K > K_*$, we can use this scheme to get a riskless profit).

Now assume that $K < K_*$.

• At time $t = 0$

– enter into a long forward contract for one share with the price $K < K_*$ written in it;
– borrow $e^{-0.04} \doteq 0.96079$ of the share with promise to return one share back at $T = 1$;
– sell this part of the share for $\$100 \times e^{-0.04} \doteq \$96.079$;
– put this money into the bank account.

• At time $t = T = 1$

– you have $K_* = 100 \times e^{-0.04} \times e^{0.02} \doteq 98.020$ ($) on your bank account;
– you buy one share for K due to the long forward and you return the share as promised.

You got the riskless profit $K_* - K > 0$. We come to a contradiction and hence we should require $K \geq K_*$ but by the first strategy we concluded that $K \leq K_*$.

Summarising, the fair forward price is $K = K_* \doteq 98.020$ ($).

7. The holder of the long position will get £1,000 at the maturity time $T = 1/2$. The present value of these £1,000 at the maturity time $T = 1/2$ is £1,000$\times e^{-0.02/2}$ and the present price of this stock on the Euro market is $S_0 = 1.200 \times 1000 \times e^{-0.02/2}$ (€). Then, the fair futures price is

$$K = 1200 \times e^{-0.02/2} \times e^{0.04/2} \doteq 1212.1 \text{ (€)}.$$

8. See Section 3.2.

9. (a) The forward price is

$$K_0 = 60.4 \times 1000 \times e^{3 \times 0.02/2} \doteq 62239 \text{ ($)}.$$

(b) The present forward price (three months later) is

$$K_1 = 47.1 \times 1000 \times e^{0.02} \doteq 48051 \ (\$).$$

The present value of the original short forward contract

$$e^{-0.02}(S_T - K_1 - [S_T - K_0]) = e^{-0.02}(K_0 - K_1)$$
$$= e^{-0.02} \times (62239 - 48051) \doteq 13907 \ (\$).$$

(c) The contract is overpriced. This creates, e.g., the following arbitrage opportunity.

- At time $t = 0$

  - enter into the short forward with the forward price $K = \$63,000$;
  - borrow $\$60.4\times$ 1,000 with the interest rate 2% p.a.;
  - using this money buy 1,000 barrels of oil and store it.

- At time $T = 3/2$

  - sell 1,000 barrels of oil for $\$63,000$ due to the short forward;
  - return $60.4\times1,000\times e^{3\times0.02/2} \doteq 62{,}239 \ (\$)$ to the bank.

Your riskless profit is $63{,}000 - 62{,}239 = 761 \ (\$)$.

10. The forward price at time $t = 0$ is equal to $K_0 = 100 \times 0.97 e^{0.035} \doteq 100.46 \ (\£)$. As it is seen from the Tables 21.1 and 21.2, the investor X receives a slightly higher profit of £19.55 than the investor Y of £19.43.

Table 21.1   Futures margins in Problem 10.

t, years	K_t	Futures: Margins
0	$97e^{0.035} \doteq 100.46$	0
3/12	$92e^{0.03\times9/12} \doteq 94.09$	$(94.09 - 100.46)e^{-0.03\times9/12} \doteq -6.23$
6/12	$95e^{0.0275\times6/12} \doteq 96.32$	$(96.32 - 94.09)e^{-0.0275\times6/12} \doteq 2.20$
9/12	$101e^{0.02\times3/12} \doteq 101.51$	$(101.51 - 96.32)e^{-0.02\times3/12} \doteq 5.16$
1	120	$120 - 101.63 = 18.37$

Table 21.2   The cash flows for Investors X and Y in Problem 10.

t, years	Forward cash flow	Futures: Cash flow
0	0	0
3/12	0	$-6.23e^{0.03/4} \doteq -6.28$
6/12	0	$(-6.28 + 2.20)e^{0.0275/4} \doteq -4.11$
9/12	0	$(-4.11 + 5.16)e^{0.02/4} \doteq 1.06$
1	$100 \times S_{12} - K_0 \doteq 19.55$	$1.06 + 18.37 = 19.43$

11. Analogous to the above problems 5(c) and 9(c).

12. In an option, only one party, the writer, takes an obligation and the other party, the holder, has a right to exercise the option. In contrast, both parties entering a forward take the obligation.

13. (i) Due to the put–call parity, we have

$$D_t = C_t - S_t + Ke^{-r(T-t)}.$$

Substituting $C_0 = \$10$ (price of the call option), $S_0 = \$125$ (price of the stock), $r = 0.03$ (continuously compounded interest rate 3%), $T = 1/2$ year (6 months), $K = \$130$ (strike price), we get

$$D_0 = \$(10 - 125 + 130 \times e^{-0.03/2}) = \$13.06.$$

(ii) At time $t = 0$, write one put for \$14, buy one call for \$10, borrow one unit of the stock and sell it for \$125 and put \$129 into the bank account, i.e. the value of the initial portfolio is 0. At the maturity time $t = 1/2$, if $S_{1/2} \leq 130$, then

$$V_{1/2} = -\left(130 - S_{1/2}\right)_+ - S_{1/2} + 129 \times e^{0.03/2} = 129 \times e^{0.03/2} - 130 \doteq 0.95(\$);$$

if $S_{1/2} > 130$, then

$$V_{1/2} = \left(S_{1/2} - 130\right)_+ - S_{1/2} + 129 \times e^{0.03/2} = 129 \times e^{0.03/2} - 130 \doteq 0.95(\$).$$

Thus, we have got a riskless profit and demonstrated an arbitrage opportunity.

14. Re-read examples and discussion in Section 5.4 and construct your own example.

15. This European-type derivative has the following payoff:

$$f(x) = \begin{cases} K_2 - K_1, & \text{if } x < K_1 \\ K_2 - x, & \text{if } K_1 \leq x \leq K_2 \\ 0, & \text{if } x > K_2. \end{cases}$$

It can be synthesised as follows: $f(x) = (K_2 - x)_+ - (K_1 - x)_+$, i.e. by buying one put with expiry at T and strike K_2 and writing one put with expiry at T and strike price K_1.

16. Let the forward price be K and the delivery date be T. The long forward contract is created by buying a European call and selling a European put which have the same strike K and the same maturity T. Indeed, the value of such a portfolio at T is $V_T = (S_T - K)_+ - (K - S_T)_+ = S_T - K$.

17. According to the put–call parity

$$C_t - D_t = S_t - Ke^{-r(T-t)}.$$

Hence, $C_t = D_t$ if and only if $K = S_t e^{r(T-t)}$, which is the forward price.

Table 21.3   The payoff of the butterfly spread.

	$S_T < 20$	$20 \le S_T < 30$	$30 \le S_T < 40$	$S_T \ge 40$
C_t^{20}	0	$S_T - 20$	$S_T - 20$	$S_T - 20$
C_t^{50}	0	0	0	$S_T - 40$
$-2C_t^{30}$	0	0	$60 - 2S_T$	$60 - 2S_T$
V_T	0	$S_T - 20$	$40 - S_T$	0

18. The value of the portfolio $V(t) = C_t^{K_1} + C_t^{K_2} - 2C_t^{K_3}$, where $C_t^{K_1}$, $C_t^{K_2}$, $C_t^{K_3}$ are the prices at time t of the corresponding calls. At the maturity we have $V_T = (S_T - 20)_+ + (S_T - 40)_+ - 2(S_T - 30)_+$ which can be represented via Table 21.3.

Note that the payoff of the butterfly spread at the maturity is always non-negative. A trader buys (takes a long position) a butterfly spread if she expects that the price of the underlying at the maturity will be between K_1 and K_2. The payoff diagram is plotted in Fig. 21.1.

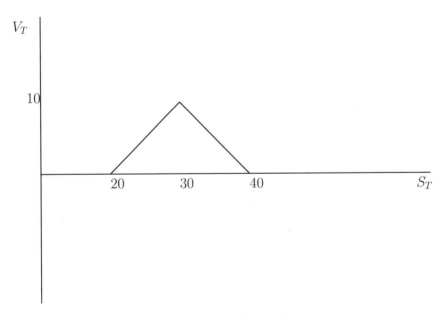

Fig. 21.1   Butterfly spread.

19. Analogous to Problem 18.

20. This is left for your own practice.

21. Due to the put–call parity:

$$C_0 = S_0 - Ke^{-rT} + D_0.$$

The price D_0 of the put is non-negative: $D_0 \geq 0$. Hence,

$$S_0 - Ke^{-rT} \leq C_0$$

as required.

22. Consider, for instance, the following portfolio. At time t buy call with the strike price K_1 for $C_t^{K_1}$, write one call with the strike price K_2 for $C_t^{K_2}$, and buy/sell bonds $B_t = C_t^{K_1} - C_t^{K_2}$ with continuously compounded interest rate r. The value of this portfolio at t is $V_t = 0$. At the maturity time $t = T$, the value of this portfolio is

$$V_T = (S_T - K_1)_+ - (S_T - K_2)_+ - B_t e^{r(T-t)},$$

i.e. if $S_T \leq K_1$, $V_T = -B_t e^{r(T-t)}$, if $K_1 < S_T \leq K_2$, $V_T = S_T - K_1 - B_t e^{r(T-t)}$, and $S_T > K_2$, $V_T = K_2 - K_1 - B_t e^{r(T-t)}$. Hence, if $B_t = C_t^{K_1} - C_t^{K_2} < 0$, we have $V_T > 0$ independently of any randomness, i.e. if $C_t^{K_1} < C_t^{K_2}$, this portfolio would offer a risk-free profit (an arbitrage opportunity). The no-arbitrage principle prohibits the existence of such prices. Therefore, the desired inequality must hold.

23. We consider two cases. If $S_t \geq K$, then obviously $D_t \geq (K - S_t)_+$ otherwise $D_t < 0$ and we get an arbitrage (please explain how). Now consider $S_t < K$ and assume that $D_t < (K - S_t)_+$. At time t form the portfolio by buying the put for D_t, buying one asset for S_t and borrowing $B = D_t + S_t$. The initial value of the portfolio $V_t = 0$. At $t = T$ its value becomes (recall that the interest rate $r = 0$):

$$V_T = (K - S_T)_+ + S_T - B = \begin{cases} K - B, & \text{if } S_T < K, \\ S_T - B, & \text{if } S_T \geq K. \end{cases}$$

Hence, $V_T \geq K - B = (K - S_t) - D_t$ which is positive since we assumed that $D_t < K - S_t$. Thus, we have come to the contradiction, the assumption $D_t < (K - S_t)_+$ is wrong, and then $D_t \geq (K - S_t)_+$ under $S_t < K$ as well.

24. A rational holder of the chooser option will prefer the option which is more valuable at time $t = s$, i.e. the payoff of the chooser option is

$$A_s = \max[D_s(K, T), C_s(K, T)],$$

where $C_s(K, T)$ and $D_s(K, T)$ are the prices at $t = s$ of the call and put with maturity time T and strike K, respectively. Due to the put–call parity relation,

$$C_t(K, T) = D_t(K, T) + S_t - Ke^{-r(T-t)}, \quad t \in [0, T],$$

and we have

$$A_s = \max[D_s(K,T), C_s(K,T)]$$
$$= max[D_s(K,T), D_s(K,T) + S_s - Ke^{-r(T-s)}]$$
$$= D_s(K,T) + (S_s - Ke^{-r(T-s)})_+$$
$$= D_s(K,T) + C_s(Ke^{-r(T-s)}, s).$$

Then the arbitrage price of the chooser option at $t = 0$ is

$$A_0 = D_0(K,T) + C_0(Ke^{-r(T-s)}, s).$$

25. You can buy a call with strike £4.5 and a put with strike £5.5 both with maturity in November. There is a guaranteed income of £1 from this portfolio, which increases if S_T lies outside the predicted range $[4.5, 5.5]$.

26. The arbitrage price is zero because the payoff of a put with zero strike is identically equal to zero.

27. (*One of possible answers.*) According to the put–call parity

$$C_t - D_t = S_t - Ke^{-r(T-t)},$$

i.e. in this case

$$C_0 - D_0 = 35 - 30e^{-0.01} \doteq 5.3 \ (\pounds).$$

An arbitrage occurs if the portfolio $C_0 - D_0$ can be bought for less than £5.3 or sold for more than £5.3. To buy this portfolio, we need to buy one call for £7 and sell one put for £1, hence this portfolio can be bought for $7 - 1 = 6$ which is larger than 5.3. To sell this portfolio, we need to sell one call for £6 and buy one put for £2 hence this portfolio can be sold for $6 - 2 = 4$ which is smaller than 5.3. Thus, there is no arbitrage in this example.

28. (An example of a possible answer.) **(i)** The pension fund can buy put options for protecting their investment because a put with strike K will guarantee that the fund can get compensation if the share price falls below the prescribed K. We also note that the fund will have to spend money on buying this protection but this premium is relatively small in comparison with possible losses if the prediction turns out to be right.

(ii) For speculation, one can also use put options which, e.g., are currently ATM, i.e. with strike K equal to the spot price S_0 of the shares. The speculator would pay a relatively small premium D_0 for buying the puts and can get substantial gains equal to $K - S_t$ if S_t goes down.

29. Compose a portfolio of the three calls, which value is $V(t) = \alpha_1 C_1(t) + \alpha_2 C_2(t) + \alpha_3 C_3(t)$. At $t = 0$ we have $V(0) = 4\alpha_1 + 3\alpha_2 + \alpha_3$. The portfolio value $V(T)$ at the maturity T is given in Table 21.4.

Table 21.4   The portfolio value $V(T)$ in Problem 29.

$S(T)$	< 100	$[100, 110)$	$[110, 120)$	≥ 120
$V(T)$	0	$\alpha_1 (S(T) - 100)$	$(\alpha_1 + \alpha_2) S(T)$ $-100\alpha_1 - 110\alpha_2$	$(\alpha_1 + \alpha_2 + \alpha_3) S(T)$ $-100\alpha_1 - 110\alpha_2$ $-120\alpha_3$

If we can find α_1, α_2, α_3 such that $V(0) = 0$ and $V(T) \geq 0$ for all possible $S(T)$ and for a range of $S(T)$ the portfolio $V(T) > 0$ then there is an arbitrage.

Then we can derive the following conditions on α_1, α_2, α_3:

$$4\alpha_1 + 3\alpha_2 + \alpha_3 = 0 \qquad (21.1)$$
$$\alpha_1 \geq 0$$
$$2\alpha_1 + \alpha_2 \geq 0$$
$$\alpha_1 + \alpha_2 + \alpha_3 \geq 0.$$

Note. The first two conditions and the last one are evident. The third condition is obtained from the requirement that

$$(\alpha_1 + \alpha_2) S(T) - 100\alpha_1 - 110\alpha_2 \geq 0 \quad \text{for} \quad all \quad 110 \leq S(T) \leq 120 \quad (21.2)$$

under $\alpha_1 \geq 0$. Consider two cases. If $\alpha_1 + \alpha_2 > 0$ then the requirement (21.2) is automatically satisfied:

$$(\alpha_1 + \alpha_2) S(T) - 100\alpha_1 - 110\alpha_2$$
$$\geq (\alpha_1 + \alpha_2) \times 110 - 100\alpha_1 - 110\alpha_2$$
$$= 10\alpha_1 \geq 0.$$

If we complimented the first two and the fourth inequalities from (21.1) with $\alpha_1 + \alpha_2 > 0$ then the corresponding system of four inequalities would be incompatible (please check). If $\alpha_1 + \alpha_2 \leq 0$ then

$$(\alpha_1 + \alpha_2) S(T) - 100\alpha_1 - 110\alpha_2$$
$$\geq (\alpha_1 + \alpha_2) \times 120 - 100\alpha_1 - 110\alpha_2$$
$$= 20\alpha_1 + 10\alpha_2,$$

which should be non-negative (otherwise there is $S(T) \in [110, 120]$ for which (21.2) does not hold). This gives us the third inequality in (21.1).

For instance, $\alpha_1 = 1$, $\alpha_2 = -1.5$, $\alpha_3 = 0.5$ satisfy the system (21.1). Indeed, for this choice of the parameters we have the portfolio value $V(T)$ shown in Table 21.5.

Thus, there is an arbitrage. The strategy to get a riskless profit consists of buying one call with $K = 100$ and $1/2$ call with $K = 120$ and selling -1.5 call with $K = 110$.

Table 21.5   The portfolio value $V(T)$ for $\alpha_1 = 1$, $\alpha_2 = -1.5$, $\alpha_3 = 0.5$ in Problem 29.

$S(T)$	< 100	$[100, 110)$	$[110, 120)$	≥ 120
$V(T)$	0	$S(T) - 100 \geq 0$	$65 - 0.5S(T) > 0$	$5 > 0$

30. Answer:

$$C_t + G_t + Ke^{-r(T-t)} = D_t + S_t.$$

31. At the maturity time the value of the first portfolio is

$$V_1(T) = 2(S_T - K)_+ = \begin{cases} 0, & S_T < K, \\ 2(S_T - K), & S_T \geq K, \end{cases}$$

and of the second

$$V_2(T) = (S_T - K + 10)_+ + (S_T - K - 10)_+$$

$$= \begin{cases} 0, & S_T < K - 10, \\ (S_T - K + 10), & K - 10 \leq S_T < K + 10, \\ 2(S_T - K), & S_T \geq K + 10. \end{cases}$$

We see that $V_2(T) \geq V_1(T)$. Hence, it is better to own the second portfolio.

32. See Section 4.2.

33. Let $t \leq T_0$. The cash flow of the fixed leg of an IRS can be replicated by selling floating rate notes for $v_n(t) = N_0 P(t, T_0)$ (see (4.18)) and buying fixed coupon bonds for $v_c(t) = (\rho\delta \sum_{i=1}^{n} P(t, T_i) + P(t, T_n)) N_0$ (see (4.14)). Then the value of the payer swap is equal to

$$\Pi_p(t) = v_n(t) - v_c(t)$$

$$= N_0 P(t, T_0) - \left(\rho\delta \sum_{i=1}^{n} P(t, T_i) + P(t, T_n) \right) N_0$$

$$= N_0 \left(P(t, T_0) - P(t, T_n) - \rho\delta \sum_{i=1}^{n} P(t, T_i) \right),$$

which coincides with (4.19) as required.

Remark 21.1. Such a replication is useful for pricing IRS when, e.g., different day count conventions[1] are used for fixed and floating legs and, in general, when the coupon dates of fixed and floating legs of an IRS do not coincide.

---

[1] You can read about day count conventions, e.g., in Hull (2003) and Filipovic (2009).

34. (i) We have

$$P(0,T) = \exp\left(-\int_0^T (0.05 + 0.01t)dt\right) = \exp\left(-0.005\,T^2 - 0.05T\right).$$

Then $P(0,1/6) \doteq 0.9916$, $P(0,5/12) \doteq 0.9785$, $P(0,2/3) \doteq 0.9651$, $P(0,11/12) \doteq 0.9512$, $P(0,7/6) \doteq 0.9369$.

(ii) Recall

$$R_{swap}(t;T_0,T_n) = \frac{P(t,T_0) - P(t,T_n)}{\delta \sum_{i=1}^n P(t,T_i)}.$$

Then $R_{swap}(0;1/6,7/6) \doteq \frac{0.9916-0.9369}{1/4\times(0.9785+0.9651+0.9512+0.9369)} = 0.0571$ (to 4 d.p.).

(iii) We have

$$P(1,T) = \exp\left(-\int_1^T (0.05 + 0.02t)dt\right) = \exp\left(-0.01T^2 - 0.05T + 0.06\right)$$

and hence $P(1,7/6) \doteq 0.9881$.

At time $t \in [T_{i-1}, T_i)$, $i = 1, \ldots, n$, the value of the receiver swap entered into at time s equals

$$\Pi_r(t) = [R_{swap}(s;T_0,T_n) - F(T_{i-1},T_i)]\,\delta N_0 P(t,T_i)$$

$$+ N_0\delta\left(R_{swap}(s;T_0,T_n) - R_{swap}(t;T_i,T_n)\right) \sum_{j=i+1}^n P(t,T_j),$$

i.e. in our case

$$\Pi_r(1) = [R_{swap}(0;1/6,7/6) - F(11/6,7/6)] \times 10^7 \times 1/4 \times P(1,7/6)$$
$$\doteq [0.0571 - 0.05] \times 10^7 \times 1/4 \times 0.9881$$
$$= 17539 \text{ (£)}.$$

35. (a) We have

$$\frac{P(0,t)}{P(0,T)} = 1 + (T - t)F(0;t,T).$$

Hence

$$P(0,T) = \frac{P(0,t)}{1 + (T - t)F(0;t,T)}.$$

Then

$$P(0, 1/12) = \frac{P(0,0)}{1 + 1/12 \times F(0; 0, 1/12)} = \frac{1}{1 + 1/12 \times 0.05}$$
$$= 0.9959 \text{ (to 4 d.p.)},$$

$$P(0, 1/6) = \frac{P(0, 1/12)}{1 + 1/12 \times F(0; 1/12, 1/6)} = \frac{0.9959}{1 + 1/12 \times 0.05}$$
$$= 0.9918 \text{ (to 4 d.p.)},$$

$$P(0, 1/4) = \frac{P(0, 1/6)}{1 + 1/12 \times F(0; 1/6, 1/4)} = \frac{0.9918}{1 + 1/12 \times 0.06}$$
$$= 0.9869 \text{ (to 4 d.p.)},$$

$$P(0, 1/3) = \frac{P(0, 1/4)}{1 + 1/12 \times F(0; 1/4, 1/3)} = \frac{0.9869}{1 + 1/12 \times 0.07}$$
$$= 0.9812 \text{ (to 4 d.p.)},$$

$$P(0, 5/12) = \frac{P(0, 1/3)}{1 + 1/12 \times F(0; 1/3, 5/12)} = \frac{0.9812}{1 + 1/12 \times 0.09}$$
$$= 0.9739 \text{ (to 4 d.p.)}.$$

(b) We have

$$R_{swap}(t; T_0, T_n) = \frac{P(t, T_0) - P(t, T_n)}{\delta \sum_{i=1}^n P(t, T_i)}.$$

Then

$$R_{swap}(0; 1/12, 5/12) = \frac{0.9959 - 0.9739}{(0.9918 + 0.9869 + 0.9812 + 0.9739)/12}$$
$$= 0.0671 \text{ (to 4 d.p.)}.$$

(c) We have

$$P(1/5, 1/4) = \frac{1}{1 + 1/20 \times F(1/5; 1/5, 1/4)} = \frac{1}{1 + 1/20 \times 0.05}$$
$$= 0.9975 \text{ (to 4 d.p.)},$$

$$P(1/5, 1/3) = \frac{P(0, 1/4)}{1 + 1/12 \times F(1/5; 1/4, 1/3)} = \frac{0.9975}{1 + 1/12 \times 0.06}$$
$$= 0.9925 \text{ (to 4 d.p.)},$$

$$P(1/5, 5/12) = \frac{P(0, 1/3)}{1 + 1/12 \times F(1/5; 1/3, 5/12)} = \frac{0.9925}{1 + 1/12 \times 0.06}$$
$$= 0.9876 \text{ (to 4 d.p.)}.$$

At time $t \in [T_{i-1}, T_i)$, $i = 1, \ldots, n$, the value of the receiver swap entered into at time s equals

$$\Pi_r(t) = [R_{swap}(s; T_0, T_n) - F(T_{i-1}, T_i)] \, \delta N_0 P(t, T_i)$$

$$+ N_0 \delta \left(R_{swap}(s; T_0, T_n) - R_{swap}(t; T_i, T_n) \right) \sum_{j=i+1}^{n} P(t, T_j),$$

i.e. in our case

$$R_{swap}(1/5; 1/4, 5/12) = \frac{P(1/5, 1/4) - P(1/5, 5/12)}{1/12 \times (P(1/5, 1/3) + P(1/5, 5/12))}$$

$$= \frac{0.9975 - 0.9876}{1/12 \times (0.9925 + 0.9876)} \doteq 0.06$$

and

$$\Pi_r(1/5) = [R_{swap}(0; 1/12, 5/12) - F(1/6, 1/4)] \times 10^8 \times 1/12 \times P(1/5, 1/4)$$

$$+ 10^8 \times 1/12 \times (R_{swap}(0; 1/12, 5/12) - R_{swap}(1/5; 1/4, 5/12))$$

$$\times (P(1/5, 1/3) + P(1/5, 5/12))$$

$$\doteq (0.0671 - 0.04) \times 10^8 \times 1/12 \times 0.9975$$

$$+ (0.0671 - 0.06) \times 10^8 \times 1/12 \times (0.9925 + 0.9876)$$

$$\doteq 342420 (\pounds).$$

36. The swap rate of the original swap entered into at time $s \leq T_0$ is $\rho = R_{swap}(s; T_0, T_n)$. The remaining payments are $[F(T_{j-1}, T_j) - R_{swap}(s; T_0, T_n)] \, \delta N_0$ at times T_j, $j = i, \ldots, n$. At time t the value of $[F(T_{i-1}, T_i) - R_{swap}(s; T_0, T_n)] \, \delta N_0$ paid at T_i is equal to $[F(T_{i-1}, T_i) - R_{swap}(s; T_0, T_n)] \, \delta N_0 P(t, T_i)$ (note that $F(T_{i-1}, T_i)$ is known at this time t). When for $T_{j-1} \geq t$, at time t the values of $[F(T_{j-1}, T_j) - R_{swap}(s; T_0, T_n)] \, \delta N_0$ paid at T_j are equal to $N_0(P(t, T_{j-1}) - P(t, T_j)) - R_{swap}(s; T_0, T_n) \delta N_0 P(t, T_j)$. Hence (by the usual convention if $i + 1 > n$ then the sum $\sum_{j=i+1}^{n}$ is assumed to be equal to zero):

$$\Pi_p(t) = [F(T_{i-1}, T_i) - R_{swap}(s; T_0, T_n)] \, \delta N_0 P(t, T_i)$$

$$+ \sum_{j=i+1}^{n} [N_0(P(t, T_{j-1}) - P(t, T_j)) - R_{swap}(s; T_0, T_n) \delta N_0 P(t, T_j)]$$

$$= [F(T_{i-1}, T_i) - R_{swap}(s; T_0, T_n)] \, \delta N_0 P(t, T_i)$$

$$+ N_0 \left(P(t, T_i) - P(t, T_n) \right) - \delta N_0 R_{swap}(s; T_0, T_n) \sum_{j=i+1}^{n} P(T_{i-1}, T_j)$$

$$= [F(T_{i-1}, T_i) - R_{swap}(s; T_0, T_n)] \, \delta N_0 P(t, T_i)$$

$$+ N_0\delta\left(R_{swap}(t; T_i, T_n) - R_{swap}(s; T_0, T_n)\right) \sum_{j=i+1}^{n} P(t, T_j).$$

Conclusion: at time $t \in [T_{i-1}, T_i)$, $i = 1, \ldots, n$, the value of the payer swap entered into at time $s \le T_0$ equals (if $t = T_{i-1}$ and $i > 0$, we assume that the payment at T_{i-1} has been already made)

$$\Pi_p(t) = [F(T_{i-1}, T_i) - R_{swap}(s; T_0, T_n)]\,\delta N_0 P(t, T_i)$$

$$+ N_0\delta\left(R_{swap}(t; T_i, T_n) - R_{swap}(s; T_0, T_n)\right) \sum_{j=i+1}^{n} P(t, T_j)$$

or equivalently

$$\Pi_p(t) = \left[\frac{1}{P(T_{i-1}, T_i)} - 1 - R_{swap}(s; T_0, T_n)\delta\right] N_0 P(t, T_i) \qquad (21.3)$$

$$+ N_0\delta\left(R_{swap}(t; T_i, T_n) - R_{swap}(s; T_0, T_n)\right) \sum_{j=i+1}^{n} P(t, T_j).$$

Note that $\Pi_p(T_{i-1})$ from (21.3) coincides with $\Pi_p(t)$ from (4.21). You may wish to do the corresponding algebraic manipulations to confirm this.

21.2 Solutions to Problems for Part II

1–2. Elementary! If not, please revise your elementary Probability knowledge using, e.g., the textbooks by Meyer (1970), Wackerly *et al.* (2007) or Chapter I of Shiryaev (1996), etc.

3. (i) Find the parameters a_n and b_n of the (I) model in this case.

- *At the first step:* $8 = 4(1 + b_1) \Rightarrow b_1 = 1$,
  $$2 = 4(1 + a_1) \Rightarrow a_1 = -1/2.$$
- *At the second step:* $16 = 8(1 + {}_1b_2) \Rightarrow {}_1b_2 = 1$,
  $$8 = 8(1 + {}_1a_2) \Rightarrow {}_1a_2 = 0;$$
  $$8 = 2(1 + {}_2b_2) \Rightarrow {}_2b_2 = 3,$$
  $$1 = 2(1 + {}_2a_2) \Rightarrow {}_2a_2 = -1/2.$$
- *At the third step:* $32 = 16(1 + {}_1b_3) \Rightarrow {}_1b_3 = 1$,
  $$16 = 16(1 + {}_1a_3) \Rightarrow {}_1a_3 = 0;$$
  $$16 = 8(1 + {}_2b_3) \Rightarrow {}_2b_3 = 1,$$
  $$4 = 8(1 + {}_2a_3) \Rightarrow {}_2a_3 = -1/2;$$
  $$4 = 1(1 + {}_4b_3) \Rightarrow {}_4b_3 = 3,$$
  $$1 = 1(1 + {}_4a_3) \Rightarrow {}_4a_3 = 0.$$

The no-arbitrage condition (Ic) is satisfied if $1 + a < e^{r\Delta t} = 5/4 < 1 + b$ at each step. Because all $a \le 0$, we have $1 + a \le 1$; minimum b is equal to 1 then $1 + b = 4/3$ which is $> 5/4$. Hence, there are no arbitrage opportunities in the market model. Note that the positiveness condition (Ib) is satisfied since all the prices are positive.

Find the implied (risk-neutral) probabilities:

$$q_1 = \frac{e^{r\Delta t} - (1 + a_1)}{b_1 - a_1} = \frac{5/4 - 1/2}{1 + 1/2} = \tfrac{1}{2}, \qquad 1 - q_1 = \tfrac{1}{2};$$

$$_1q_2 = \frac{e^{r\Delta t} - (1 + {}_1a_2)}{{}_1b_2 - {}_1a_2} = \frac{5/4 - 1}{1 - 0} = \tfrac{1}{4}, \qquad 1 - {}_1q_2 = \tfrac{3}{4};$$

$$_2q_2 = \frac{e^{r\Delta t} - (1 + {}_2a_2)}{{}_2b_2 - {}_2a_2} = \frac{5/4 - 1/2}{3 + 1/2} = \tfrac{3}{14} \doteq 0.2142, \qquad 1 - {}_2q_2 = \tfrac{11}{14};$$

$$_1q_3 = \frac{e^{r\Delta t} - (1 + {}_1a_3)}{{}_1b_3 - {}_1a_3} = \frac{5/4 - 1}{1 - 0} = \tfrac{1}{4}, \qquad 1 - {}_1q_3 = \tfrac{3}{4};$$

$$_2q_3 = \frac{e^{r\Delta t} - (1 + {}_2a_3)}{{}_2b_3 - {}_2a_3} = \frac{5/4 - 1/2}{1 + 1/2} = \tfrac{1}{2}, \qquad 1 - {}_2q_3 = \tfrac{1}{2};$$

$$_3q_3 = \frac{e^{r\Delta t} - (1 + {}_3a_3)}{{}_3b_3 - {}_3a_3} = \frac{5/4 - 1}{3 + 0} = \tfrac{1}{12}, \qquad 1 - {}_3q_3 = \tfrac{11}{12}.$$

(ii) (a) The writer's obligation at $t = 3\Delta t$: $f(S_3) = (16 - S_3)_+$, i.e. $_1f_3 = 0$, $_2f_3 = 0$, $_3f_3 = 12$, $_4f_3 = 15$.

We have

$$_1V_2 = e^{-r\Delta t} E_Q(f(S_3)|\mathcal{A}_2)(UU)$$
$$= e^{-r\Delta t}\left[{}_1q_3 \times {}_1f_3 + (1 - {}_1q_3) \times_2 f_3\right] = 0 \ (\pounds);$$
$$_2V_2 = e^{-r\Delta t} E_Q(f(S_3)|\mathcal{A}_2)(UD) = e^{-r\Delta t} E_Q(f(S_3)|\mathcal{A}_2)(DU)$$
$$= e^{-r\Delta t}\left[{}_2q_3 \times {}_2f_3 + (1 - {}_2q_3) \times_3 f_3\right]$$
$$= \tfrac{4}{5}(\tfrac{1}{2} \times 0 + \tfrac{1}{2} \times 12) = \tfrac{24}{5} \ (\pounds);$$
$$_3V_2 = e^{-r\Delta t} E_Q(f(S_3)|\mathcal{A}_2)(DD)$$
$$= e^{-r\Delta t}\left[{}_3q_3 \times {}_3f_3 + (1 - {}_3q_3) \times_4 f_3\right]$$
$$= \tfrac{4}{5}(\tfrac{1}{12} \times 12 + \tfrac{11}{12} \times 15) = \tfrac{59}{5} \ (\pounds);$$

$$_1V_1 = e^{-r\Delta t} E_Q(V_2|\mathcal{A}_1)(U) = e^{-r\Delta t}\left[{}_1q_2 \times {}_1V_2 + (1 - {}_1q_2) \times {}_2V_2\right]$$
$$\doteq \tfrac{4}{5}(\tfrac{1}{4} \times 0 + \tfrac{3}{4} \times \tfrac{24}{5}) = \tfrac{72}{25} = 2.88 \ (\pounds);$$
$$_2V_1 = e^{-r\Delta t} E_Q(V_2|\mathcal{A}_1)(D) = e^{-r\Delta t}\left[{}_2q_2 \times {}_2V_2 + (1 - {}_2q_2) \times {}_3V_2\right]$$
$$\doteq \tfrac{4}{5}(\tfrac{3}{14} \times \tfrac{24}{5} + \tfrac{11}{14} \times \tfrac{59}{5}) = \tfrac{206}{25} = 8.24 \ (\pounds);$$

$$x = V_0 = e^{-r\Delta t} E_Q V_1 = e^{-r\Delta t}\left[q_1 \times {}_1V_1 + (1 - q_1) \times_2 V_1\right]$$
$$\doteq \tfrac{4}{5}(\tfrac{1}{2} \times \tfrac{72}{25} + \tfrac{1}{2} \times \tfrac{206}{25}) = \tfrac{556}{125} \doteq 4.45 \ (\pounds).$$

Answer. The fair premium for this option paid at $t = 0$ is £4.45.

(b) Because we assume that $S_1 = {}_1S_1 = \pounds 8$ is realised, we need to find the portfolio $({}_1\phi_1,\ {}_1\psi_1)$. We have

$$
{}_1\phi_1 = \frac{{}_1V_2 - {}_2V_2}{{}_1S_2 - {}_2S_2} = \frac{0 - \frac{24}{5}}{16 - 8} = -\frac{3}{5},
$$

$$
{}_1\psi_1 = {}_1V_1 - {}_1S_1\,{}_1\phi_1 = \frac{72}{25} + 8 \times \tfrac{3}{5} = \frac{192}{25} = 7.68.
$$

Answer. If $S_1 = {}_1S_1 = £8$ is realised at time $t = \Delta t$, the writer forms the portfolio consisting of $-\dfrac{3}{5}$ unit of stock (i.e. borrowed $\dfrac{3}{5}$ stock) and £7.68 of money on the bank account.

4. The payoff for the American-style arithmetic-average *Asian put* option with strike $K = £16$ is

$$
f_n = \left(16 - \frac{1}{n+1}\sum_{i=0}^{n} S_i\right)_+ .
$$

We have $f_N(UUU) = 1$, $f_N(UUD) = 5$, $f_N(UDU) = 7$, $f_N(UDD) = 10$, $f_N(DUU) = 8.5$, $f_N(DUD) = 11.2$, $f_N(DDU) = 13.25$, $f_N(DDD) = 14$.

Then

$$
\begin{aligned}
{}_1U_2 &= e^{-r_2\Delta t}E_Q(f(S_3)|\mathcal{A}_2)(UU)\\
&= e^{-r\Delta t}\left[{}_1q_3 \times f_N(UUU) + (1 -{}_1 q_3) \times f_N(UUD)\right]\\
&= \tfrac{4}{5}(\tfrac{1}{4} \times 1 + \tfrac{3}{4} \times 5) = \tfrac{16}{5}\ (£),\\
{}_2U_2 &= e^{-r\Delta t}E_Q(f(S_3)|\mathcal{A}_2)(UD)\\
&= e^{-r\Delta t}\left[{}_2q_3 \times f_N(UDU) + (1 -{}_2 q_3) \times f_N(UDD)\right]\\
&= \tfrac{4}{5}(\tfrac{1}{2} \times 7 + \tfrac{1}{2} \times 10) = \tfrac{34}{5}\ (£),\\
{}_3U_2 &= e^{-r\Delta t}E_Q(f(S_3)|\mathcal{A}_2)(DU)\\
&= e^{-r\Delta t}\left[{}_2q_3 \times f_N(DUU) + (1 -{}_2 q_3) \times f_N(DUD)\right]\\
&= \tfrac{4}{5}(\tfrac{1}{2} \times 8.5 + \tfrac{1}{2} \times 11.2) = 7.88\ (£),\\
{}_4U_2 &= e^{-r\Delta t}E_Q(f(S_3)|\mathcal{A}_2)(DD)\\
&= e^{-r\Delta t}\left[{}_3q_3 \times f_N(DDU) + (1 -{}_3 q_3) \times f_N(DDD)\right]\\
&= \tfrac{4}{5}(\tfrac{1}{12} \times 13.25 + \tfrac{11}{12} \times 14) = 11.15\ (£).
\end{aligned}
$$

We have ${}_1W_2 = f_2(UU) = 16 - 28/3 = \tfrac{20}{3}\ (£)$, ${}_2W_2 = f_2(UD) = 16 - 20/3 = \tfrac{28}{3}\ (£)$, ${}_3W_2 = f_2(DU) = 16 - 14/3 = \tfrac{34}{3}\ (£)$, ${}_4W_2 = f_2(DD) = 16 - 7/3 = \tfrac{41}{3} = 13.667\ (£)$.

Then

$$
\begin{aligned}
{}_1V_2 &= \max({}_1U_2, {}_1W_2) = \tfrac{20}{3}\ (£), \quad {}_2V_2 = \tfrac{28}{3}\ (£),\\
{}_3V_2 &= \tfrac{34}{3}\ (£),\ {}_4V_2 = \tfrac{41}{3}\ (£)
\end{aligned}
$$

and

$$
\begin{aligned}
{}_1U_1 &= e^{-r\Delta t}E_Q(V_2|\mathcal{A}_1)(U) = e^{-r\Delta t}\left[{}_1q_2 \times {}_1V_2 + (1 -{}_1 q_2) \times {}_2V_2\right]\\
&= \tfrac{4}{5}(\tfrac{1}{4} \times \tfrac{20}{3} + \tfrac{3}{4} \times \tfrac{28}{3}) = \tfrac{104}{15} \doteq 6.93\ (£),\\
{}_2U_2 &= e^{-r\Delta t}E_Q(V_2|\mathcal{A}_1)(D) = e^{-r\Delta t}\left[{}_2q_2 \times {}_3V_2 + (1 -{}_2 q_2) \times {}_4V_2\right]
\end{aligned}
$$

$$= \tfrac{4}{5}(\tfrac{3}{14} \times \tfrac{34}{3} + \tfrac{11}{14} \times \tfrac{41}{3}) = \tfrac{158}{15} \doteq 10.53 \ (\pounds).$$

We have $_1W_1 = f_1(U) = 10 \ (\pounds)$, $_2W_1 = f_1(D) = 13 \ (\pounds)$.

Then

$$_1V_1 = \max(_1U_1, _1W_1) = 10 \ (\pounds), \quad _2V_2 = 13 \ (\pounds)$$

and

$$V_0 = e^{-r\Delta t} E_Q \ V_1 = e^{-r\Delta t} \left[q_1 \times \ _1V_1 + (1 - q_1) \times_2 V_1 \right]$$

$$\doteq \tfrac{4}{5}(\tfrac{1}{2} \times 10 + \tfrac{1}{2} \times 13) = \tfrac{46}{5} = 9.2 \text{ (it's } < W_0 = 12).$$

Thus, the price for this American option at time $n = 0$ is 12 $(\pounds)$.

Here, the optimal exercise time τ is a random variable taking values in the set $\{0, 1, 2, 3, \infty\}$. According to the American option pricing algorithm, the optimal exercise time τ is found as $\tau_0 = \inf_{0 \le m \le 3}\{m : V_m = f(S_m)$ and $f(S_m) > 0\}$. We have here $\tau(\omega) = 0$ on all ω.

5. We have:

$$Z(\text{`}UUU\text{'}) = (1/2 \times 1/4 \times 1/4)/(1/8) = \tfrac{1}{4};$$
$$Z(\text{`}UUD\text{'}) = (1/2 \times 1/4 \times 3/4)/(1/8) = \tfrac{3}{4};$$
$$Z(\text{`}UDU\text{'}) = (1/2 \times 3/4 \times 1/2)/(1/8) = \tfrac{3}{2};$$
$$Z(\text{`}UDD\text{'}) = (1/2 \times 3/4 \times 1/2)/(1/8) = \tfrac{3}{2};$$
$$Z(\text{`}DUU\text{'}) = (1/2 \times 3/14 \times 1/2)/(1/8) = \tfrac{3}{7};$$
$$Z(\text{`}DUD\text{'}) = (1/2 \times 3/14 \times 1/2)/(1/8) = \tfrac{3}{7};$$
$$Z(\text{`}DDU\text{'}) = (1/2 \times 11/14 \times 1/12)/(1/8) = \tfrac{11}{42};$$
$$Z(\text{`}DDU\text{'}) = (1/2 \times 11/14 \times 11/12)/(1/8) = \tfrac{121}{42}.$$

6. (i) Find the parameters a_n and b_n of the (I) model in this case.

- *At the first step*: $20 = 10(1 + b_1) \Rightarrow b_1 = 1$,
$$5 = 10(1 + a_1) \Rightarrow a_1 = -1/2.$$

- *At the second step*: $30 = 20(1 + {}_1b_2) \Rightarrow {}_1b_2 = 1/2$,
$$15 = 20(1 + {}_1a_2) \Rightarrow {}_1a_2 = -1/4;$$
$$15 = 5(1 + {}_2b_2) \Rightarrow {}_2b_2 = 2;$$
$$5 = 5(1 + {}_2a_2) \Rightarrow {}_2a_2 = 0.$$

- *At the third step*: $40 = 30(1 + {}_1b_3) \Rightarrow {}_1b_3 = 1/3$,
$$20 = 30(1 +_1 a_3) \Rightarrow {}_1a_3 = -1/3;$$
$$20 = 15(1 + {}_2b_3) \Rightarrow {}_2b_3 = 1/3;$$
$$10 = 15(1 +_2 a_3) \Rightarrow {}_2a_3 = -1/3;$$
$$10 = 5(1 + {}_4b_3) \Rightarrow {}_4b_3 = 1;$$
$$5 = 5(1 +_4 a_3) \Rightarrow {}_4a_3 = 0.$$

The no-arbitrage condition (Ic) is satisfied if $1 + a < e^{r\Delta t} = 5/4 < 1 + b$ at each step. Because all $a \le 0$, we have $1 + a \le 1$; minimum b is equal to $1/3$ then $1 + b = 4/3$ which is $> 5/4$. Hence, there are no arbitrage opportunities in the market model. Note that the condition (Ib) is also satisfied.

Find the implied (risk-neutral) probabilities:

$$q_1 = \frac{e^{r\Delta t} - (1 + a_1)}{b_1 - a_1} = \frac{5/4 - 1/2}{1 + 1/2} = \frac{1}{2}, \quad 1 - q_1 = \frac{1}{2};$$

$$_1q_2 = \frac{e^{r\Delta t} - (1 + \,_1a_2)}{_1b_2 - \,_1a_2} = \frac{5/4 - 3/4}{1/2 + 1/4} = \frac{2}{3}, \quad 1 - \,_1q_2 = \frac{1}{3};$$

$$_2q_2 = \frac{e^{r\Delta t} - (1 + \,_2a_2)}{_2b_2 - \,_2a_2} = \frac{5/4 - 1}{2 + 0} = \frac{1}{8} = 0.125, \quad 1 - \,_2q_2 = \frac{7}{8} = 0.875;$$

$$_1q_3 = \frac{e^{r\Delta t} - (1 + \,_1a_3)}{_1b_3 - \,_1a_3} = \frac{5/4 - 2/3}{1/3 + 1/3} = \frac{7}{8} \doteq 0.875, \quad 1 - \,_1q_3 = \frac{1}{8};$$

$$_2q_3 = \frac{e^{r\Delta t} - (1 + \,_2a_3)}{_2b_3 - \,_2a_3} = \frac{5/4 - 2/3}{1/3 + 1/3} = \frac{7}{8} = 0.875,$$

$$1 - \,_2q_3 = \frac{1}{8} = 0.125;$$

$$_3q_3 = \frac{e^{r\Delta t} - (1 + \,_3a_3)}{_3b_3 - \,_3a_3} = \frac{5/4 - 1}{1 + 0} = \frac{1}{4}, \quad 1 - \,_3q_3 = \frac{3}{4}.$$

(ii) (a) The writer's obligation at $t = 3\Delta t$: $f(S_3) = (S_3 - 10)_+$, i.e. $_1f_3 = 30$, $_2f_3 = 10$, $_3f_3 = 0$, $_4f_3 = 0$.

We have

$$_1V_2 = e^{-r\Delta t} E_Q(f(S_3)|\mathcal{A}_2)(UU)$$
$$= e^{-r\Delta t} \left[_1q_3 \times \,_1f_3 + (1 -_1 q_3) \times_2 f_3\right]$$
$$= \tfrac{4}{5}(\tfrac{7}{8} \times 30 + \tfrac{1}{8} \times 10) = 22 \ (\text{£}),$$

$$_2V_2 = e^{-r\Delta t} E_Q(f(S_3)|\mathcal{A}_2)(UD) = e^{-r\Delta t} E_Q(f(S_3)|\mathcal{A}_2)(DU)$$
$$= e^{-r\Delta t} \left[_2q_3 \times \,_2f_3 + (1 -_2 q_3) \times_3 f_3\right]$$
$$= \tfrac{4}{5}(\tfrac{7}{8} \times 10 + \tfrac{1}{8} \times 0) = 7 \ (\text{£});$$

$$_3V_2 = e^{-r\Delta t} E_Q(f(S_3)|\mathcal{A}_2)(DD)$$
$$= e^{-r\Delta t} \left[_3q_3 \times \,_3f_3 + (1 -_3 q_3) \times_4 f_3\right] = 0 \ (\text{£});$$

$$_1V_1 = e^{-r\Delta t} E_Q(V_2|\mathcal{A}_1)(U) = e^{-r\Delta t} \left[_1q_2 \times \,_1V_2 + (1 -_1 q_2) \times \,_2V_2\right]$$
$$\doteq \tfrac{4}{5}(\tfrac{2}{3} \times 22 + \tfrac{1}{3} \times 7) = \tfrac{68}{5} = 13.6 \ (\text{£});$$

$$_2V_1 = e^{-r\Delta t} E_Q(V_2|\mathcal{A}_1)(D) = e^{-r\Delta t} \left[_2q_2 \times \,_2V_2 + (1 -_2 q_2) \times \,_3V_2\right]$$
$$\doteq \tfrac{4}{5}(\tfrac{1}{8} \times 7 + \tfrac{7}{8} \times 0) = 0.7 \ (\text{£});$$

$$x = V_0 = e^{-r\Delta t} E_Q \, V_1 = e^{-r\Delta t} \left[q_1 \times \,_1V_1 + (1 - q_1) \times_2 V_1\right]$$
$$\doteq \tfrac{4}{5}(\tfrac{1}{2} \times 13.6 + \tfrac{1}{2} \times 0.7) = 5.72 \ (\text{£}).$$

Answer. The fair premium for this option paid at $t = 0$ is £5.72.

(b) Since we assume that $S_2 = \,_2S_2 = £15$ is realised, we need to find the portfolio $(_2\phi_2, \,_2\psi_2)$. We have

$$_2\phi_2 = \frac{_2V_3 - \,_3V_3}{_2S_3 -_3 S_3} = \frac{10 - 0}{20 - 10} = 1, \quad _2\psi_2 =_2 V_2 - \,_2S_2 \,_2\phi_2 = -8.$$

Answer. If $S_2 = \,_2S_2 = £15$ is realised at time $t = \Delta t$, the writer forms the portfolio consisting of 1 unit of stock and borrowed £8.

(c) According to the put–call parity

$$D_t = C_t - S_t + Ke^{-r(T-t)}.$$

Then the price of the corresponding put is

$$D_0 \doteq 5.72 - 10 + 10 \times \frac{4}{5} = 3.72 \ (\pounds).$$

(d) A European-type arithmetic-average Asian call option has the payoff which depends on the path:

$$f_N = \left(\frac{1}{N+1} \sum_{n=0}^{N} S_n - K \right)_+$$

and which can be exercised at the maturity time N only. Then, in the case of this tree, we have

$f_N(UUU) = (10 + 20 + 30 + 40)/4 - 10 = 15,$
$f_N(UUD) = (10 + 20 + 30 + 20)/4 - 10 = 10,$
$f_N(UDU) = (10 + 20 + 15 + 20)/4 - 10 = \frac{25}{4},$
$f_N(UDD) = (10 + 20 + 15 + 10)/4 - 10 = 15/4,$
$f_N(DUU) = (10 + 5 + 15 + 20)/4 - 10 = 5/2,$
$f_N(DUD) = (10 + 5 + 15 + 10)/4 - 10 = 0,$
$f_N(DDU) = ((10 + 5 + 5 + 10)/4 - 10)_+ = (-\frac{5}{2})_+ = 0,$
$f_N(DDD) = ((10 + 5 + 5 + 5)/4 - 10)_+ = (-\frac{15}{4})_+ = 0.$

7. Analogous to Problem 4. The answer is: the price for this American option at time $n = 0$ is 12.747 ($\pounds$) and the optimal exercise time τ is a random variable taking values in the set $\{0, 1, 2, 3, \infty\}$, which is given in Table 21.6.

Table 21.6   The optimal exercise time τ in Problem 7.

ω	DDD	DDU	DUD	DUU	UDD	UDU	UUD	UUU
$\tau(\omega)$	1	1	1	1	2	2	3	3

8. Analogous to Problems 3 and 6. The answer on (b)(1) is that the fair premium for this option paid at $t = 0$ is £26.82.

9. Analogous to Problems 4 and 7. The answer is: $_1V_2 = \max(_1U_2, _1W_2) \doteq$ 44.6 ($\pounds$) – on path UU; $_2V_2 = 60$ ($\pounds$) – on path UD; $_3V_2 \doteq 24.78$ ($\pounds$) – on path DU; $_4V_2 = 60$ ($\pounds$) – on path DD.

10. In this case a put satisfies the inequality (12.12) for all t and it would be foolish to exercise the American put before the maturity time (see the arguments after (12.12)).

11. (a) The smallest is $\mathcal{A}_7$, i.e. once the coin has been tossed seven times, we can decide whether the event A has occurred or not. However, at $n = 6$

it is not always possible to tell whether A has occurred or not: for instance, if the outcomes of the first six tosses are $\{T, T, H, H, H, H\}$, then we cannot say whether event A occurred or not and we have to toss the coin one more time to be able to make a judgement about A.

(b) Analogous to (a), the answer is $\mathcal{A}_{10}$.

(c) Analogous to (a) and (b), the answer is $\mathcal{A}_N$.

(d) We have $G = \emptyset$ which belongs to every $\mathcal{A}_0$, $n = 0, 1, 2, \dots$. Hence, the answer is $\mathcal{A}_0$.

12. See the recommendation for Problem 1 earlier in this section.

13. It is clear (why?) that $E|\eta_n| < \infty$ for every n. We have $E\left[\eta_{n+1} | \mathcal{A}_n\right] = E\left[\sum_{i=1}^{n+1} \zeta_i \,\middle|\, \mathcal{A}_n\right] = \sum_{i=1}^{n+1} E\left(\zeta_i | \mathcal{A}_n\right)$. Because ζ_i, $i \leq n$, are measurable with respect to $\mathcal{A}_n$, $E\left(\zeta_i | \mathcal{A}_n\right) = \zeta_i$ for $i \leq n$. Hence, $E\left[\eta_{n+1} | \mathcal{A}_n\right] = \sum_{i=1}^{n} \zeta_i + E\left(\zeta_{n+1} | \mathcal{A}_n\right)$. Due to the assumptions, $P(\zeta_{n+1} = \pm 1 | \zeta_1, \dots, \zeta_n) = 1/2$. Then $E\left(\zeta_{n+1} | \mathcal{A}_n\right) = \frac{1}{2} \times 1 + \frac{1}{2} \times (-1) = 0$. Thus, $E\left[\eta_{n+1} | \mathcal{A}_n\right] = \sum_{i=1}^{n} \zeta_i = \eta_n$, i.e. η_n is a martingale with respect to the given probability measure.

14. The property $E|S_n| < \infty$ is obviously satisfied. Because ζ_{n+1} is independent of $\mathcal{A}_n$ and η_n is $\mathcal{A}_n$- measurable, we have

$$
\begin{aligned}
E(\theta_{n+1} | \mathcal{A}_n) &= E((-1)^{n+1} \cos(\pi \eta_{n+1}) | \mathcal{A}_n) \\
&= E((-1)^{n+1} \cos(\pi(\eta_n + \zeta_{n+1})) | \mathcal{A}_n) \\
&= (-1)^{n+1} E(\cos(\pi \eta_n) \cos(\pi \zeta_{n+1})) | \mathcal{A}_n) \\
&\quad - (-1)^{n+1} E(\sin(\pi \eta_n) \sin(\pi \zeta_{n+1})) | \mathcal{A}_n) \\
&= (-1)^{n+1} \cos(\pi \eta_n) E(\cos(\pi \zeta_{n+1})) \\
&\quad - (-1)^{n+1} \sin(\pi \eta_n) E(\sin(\pi \zeta_{n+1})) \\
&= (-1)^{n+1} \cos(\pi \eta_n) \frac{1}{2} (\cos(-\pi) + \cos(\pi)) \\
&\quad - (-1)^{n+1} \sin(\pi \eta_n) \frac{1}{2} (\sin(-\pi) + \sin(\pi)) \\
&= (-1)^n \cos(\pi \eta_n) = \theta_n
\end{aligned}
$$

as required.

15. Analogous to the above Problems 13 and 14.

16. Since $\tau \leq N$, we have

$$
\eta_\tau = \eta_0 + \sum_{i=0}^{\tau \wedge (N-1)} (\eta_{i+1} - \eta_i) = \eta_0 + \sum_{i=0}^{N-1} I_{\tau > i}(\omega) \cdot (\eta_{i+1} - \eta_i).
$$

Using the fact that the event $\{\tau(\omega) > i\} \in \mathcal{A}_i$ and hence $I_{\tau > i}(\omega)$ is $\mathcal{A}_i$-measurable and that η_i is a martingale, we obtain

$$E\eta_\tau = E\eta_0 + \sum_{i=0}^{N-1} E\left\{I_{\tau > i}(\omega) \cdot (\eta_{i+1} - \eta_i)\right\}$$

$$= E\eta_0 + \sum_{i=0}^{N-1} E\left\{E\left[I_{\tau > i}(\omega) \cdot (\eta_{i+1} - \eta_i)\,|\mathcal{A}_i\right]\right\}$$

$$= E\eta_0 + \sum_{i=0}^{N-1} E\left\{I_{\tau > i}(\omega) \cdot E\left[(\eta_{i+1} - \eta_i)\,|\mathcal{A}_i\right]\right\}$$

$$= E\eta_0$$

as required.

17. The property $E|S_n| < \infty$ is obviously satisfied. Note that the conditions $1 - \mu - \sigma \geq 0$ and $S_0 > 0$ ensure that $S_n > 0$ for all n. We have

$$E(S_{n+1}|\mathcal{A}_n) = E(S_n - \mu S_n + \sigma S_n \xi_{n+1}|\mathcal{A}_n)$$

$$= S_n - \mu S_n + \sigma S_n E(\xi_{n+1}|\mathcal{A}_n)$$

$$= S_n - \mu S_n + \sigma S_n E(\xi_{n+1})$$

$$= S_n(1 - \mu) \leq S_n$$

and hence we showed that the process S_n is a supermartingale as required. For $\mu = 0$, we get $E(S_{n+1}|\mathcal{A}_n) = S_n$, i.e. for $\mu = 0$ this process is a martingale.

18. Definition 9.6 is equivalent to saying that $E_Q(\xi|\mathcal{A}_m)$ is an $\mathcal{A}_m$-measurable random variable such that

$$E_Q(I_A \cdot E_Q(\xi|\mathcal{A}_m)) = E_Q(I_A \cdot \xi) \tag{21.4}$$

for every $A \in \mathcal{A}_m$. Hence, to prove the equality (9.17), it is sufficient to show that

$$E_Q\left(I_A \cdot \frac{1}{Z_m} E_P(Z_n \zeta_n|\mathcal{A}_m)\right) = E_Q(I_A \cdot \zeta_n)$$

for any $A \in \mathcal{A}_m$.

Introduce the $\mathcal{A}_m$-measurable random variable

$$\Xi_m := \frac{1}{Z_m} E_P(Z_n \zeta_n|\mathcal{A}_m).$$

We have for any $A \in \mathcal{A}_m \subseteq \mathcal{A}_n$:

$$E_Q(I_A \cdot \Xi_m) \overset{(9.16)}{=} E_P(I_A \cdot \Xi_m \cdot Z_m) = E_P(I_A \cdot E_P(Z_n \zeta_n|\mathcal{A}_m))$$

$$\overset{(21.4)}{=} E_P(I_A Z_n \zeta_n) \overset{(9.16)}{=} E_Q(I_A \cdot \zeta_n).$$

19. Analogous to the proof of (9.14).

20. (*One of possible answers*). We know (see Example 11.2) that if a single-period market consists of M tradable assets and at the end of the time period the market is in one of L possible states then this market is complete if and only if the rank of the matrix of asset prices (which we denote by D) is L. In the given example a single period of the model we have two tradable assets ($M = 2$) and four possible states ($L = 4$), and the rank of D is ≤ 2. So, this market model is incomplete.

21. (*One of possible solutions*). **(a)** Due to the first fundamental theorem of asset pricing, the model is arbitrage free if and only if there is an EMM. Let us find such measures. We need to find Q such that

$$E_Q\left(e^{-r(n+1)\Delta t}S_{n+1}|\mathcal{A}_n\right) = e^{-rn\Delta t}S_n.$$

We have

$$E_Q\left(e^{-r(n+1)\Delta t}S_{n+1}|\mathcal{A}_n\right) = e^{-r(n+1)\Delta t}E_Q\left(S_n + S_n\xi_{n+1}|\mathcal{A}_n\right)$$
$$= e^{-r(n+1)\Delta t}S_n(1 + E_Q\left(\xi_{n+1}|\mathcal{A}_n\right))$$
$$= e^{-r(n+1)\Delta t}S_n(1 + cq_1 + bq_2 + a(1 - q_1 - q_2))$$

for some probabilities $Q(\xi_n = c) = q_1$, $Q(\xi_n = b) = q_2$, $Q(\xi_n = a) = q_3 = 1 - q_1 - q_2$ with $0 < q_1, q_2 < 1$, $q_1 + q_2 < 1$ (these Q-probabilities should be non-zero because the corresponding P-probabilities are assumed to be non-zero). Then the discounted price is a Q-martingale if (the condition $E_Q|S_n| < \infty$ is satisfied automatically here):

$$1 + (c - a)q_1 + (b - a)q_2 + a = e^{r\Delta t}.$$

Now let $q = q_1$ be a parameter such that $0 < q < 1$ and substitute the given values of a, b, c and r:

$$q_2 = \frac{e^{r\Delta t} - (1 + a) - q(c - a)}{b - a}$$
$$= \frac{1.05 - 0.5 - 1.5q}{1}$$
$$= 0.55 - 1.5q.$$

The conditions $0 < q_1, q_2, q_1 + q_2 < 1$ imply that

$$0 < q < 1$$
$$0 < 0.55 - 1.5q < 1$$
$$0 < 0.55 - 0.5q < 1.$$

Hence, for q

$$0 < q < 11/30,$$

we obtain the one-parametric family of EMMs

$$Q(\xi_n = 1) = q, \ Q(\xi_n = 0.5) = 0.55 - 1.5q, \ Q(\xi_n = -1) = 0.5q - 0.45.$$

Thus, this market is arbitrage free (due to the first fundamental theorem of asset pricing) and incomplete (due to the second fundamental theorem of asset pricing).

(b) The payoff of this call is $f(S_1) = (S_1 - 5)_+$ and hence $f(_1S_1) = 15$, $f(_2S_1) = 10$, $f(_3S_1) = 0$. The portfolio of $\phi_0 = 1$ stock and $\psi = -5/1.05$ money on the bank account at time 0 replicates this claim at time $n = 1$:

$$20 \cdot 1 - \frac{5}{1.05}1.05 = 15$$

$$15 \cdot 1 - \frac{5}{1.05}1.05 = 10$$

$$5 \cdot 1 - \frac{5}{1.05}1.05 = 0.$$

According to Proposition 9.3 from Section 9.5, this option has the unique arbitrage price (despite the market being incomplete).

22. Analogous to the above Problems 20 and 21.

23. It is for your self-study.

24. Hint: apply Jensen's inequality from Problem 23 to $\varphi(x) = \ln x$.

25. Using elementary Probability and Calculus, we obtain $x = \sigma_1 \rho / \sigma_2$.

21.3   Solutions to Problems for Part III

1–3. See the recommendation for Problem 1 in Section 21.2.

4. We have

$$S_n(1 - \mu \Delta t/2) = S_{n-1}(1 + \mu \Delta t/2 + \sigma \sqrt{\Delta t} \xi_n).$$

Let

$$\zeta_n := \ln \frac{S_n}{S_{n-1}} = \ln \left(\frac{1 + \mu \Delta t/2 + \sigma \sqrt{\Delta t} \xi_n}{1 - \mu \Delta t/2} \right)$$

$$= \ln(1 + \mu \Delta t/2 + \sigma \sqrt{\Delta t} \xi_n) - \ln(1 - \mu \Delta t/2).$$

Due to the assumptions made, ξ_n, $n = 1, \ldots, N$, are independent, consequently $\dfrac{S_n}{S_{n-1}}$ are independent and, hence, ζ_n are independent under the measure P.

We obtain

$$E_P \zeta_n = \frac{1}{2} \ln(1 + \mu \Delta t/2 + \sigma \sqrt{\Delta t}) + \frac{1}{2} \ln(1 + \mu \Delta t/2 - \sigma \sqrt{\Delta t})$$
$$- \ln(1 - \mu \Delta t/2)$$
$$= \left(\mu - \frac{\sigma^2}{2} \right) \Delta t + O((\Delta t)^2),$$

and

$$Var_P \zeta_n = \sigma^2 \Delta t + O((\Delta t)^2).$$

We have

$$\log \frac{S_n}{S_0} = \log \left(\frac{S_1}{S_0} \frac{S_2}{S_1} \cdots \frac{S_{n-1}}{S_{n-2}} \frac{S_n}{S_{n-1}} \right) = \sum_{i=1}^{n} \log \frac{S_i}{S_{i-1}} = \sum_{i=1}^{n} \zeta_i.$$

Let $\eta_n := \dfrac{\zeta_n - E\zeta_n}{\sqrt{Var\zeta_n}}$. Then

$$\log \frac{S_N}{S_0} = E \sum_{i=1}^{N} \zeta_i + \sum_{i=1}^{N} \frac{\zeta_i - E\zeta_i}{\sqrt{Var\zeta_i}} \sqrt{Var\zeta_i}$$

$$= \left(\mu - \frac{\sigma^2}{2} \right) T + O(\Delta t) + \left[\sigma \sqrt{T} + O(\Delta t) \right] \frac{\sum_{i=1}^{N} \eta_i}{\sqrt{N}}.$$

Since $O(\Delta t) \to 0$ as $\Delta t \to 0$ and, due to CLT, we get

$$\frac{\sum_{i=1}^{N} \eta_i}{\sqrt{N}} \Longrightarrow \tilde{\theta} \sim \mathcal{N}(0,1) \quad \text{as} \quad \Delta t \to 0,$$

we obtain that

$$\log \frac{S_N}{S_0} \Longrightarrow \theta \sim \mathcal{N}\left(\left(\mu - \frac{\sigma^2}{2} \right) T, \sigma^2 T \right) \quad \text{as} \quad \Delta t \to 0.$$

Hence

$$S_N \Longrightarrow S_0 e^\theta \quad \text{with} \quad \theta \sim \mathcal{N}\left(\left(\mu - \frac{\sigma^2}{2} \right) T, \sigma^2 T \right) \quad \text{as} \quad \Delta t \to 0.$$

5. The solution is analogous to the solution of the previous problem and also to similar problems in Sections 16.1 and 16.2. The answer is

$$S_N \Longrightarrow S_0 e^\theta \quad \text{with} \quad \theta \sim \mathcal{N}\left(\frac{\sigma^2}{2} T, \sigma^2 T \right) \quad \text{as} \quad \Delta t \to 0.$$

6. We have

$$E\left[\left(S_0\exp\left(-\frac{\sigma^2}{2}T+\sigma\sqrt{T}\theta\right)-Ke^{-rT}\right)_+\right]$$

$$=\frac{1}{\sqrt{2\pi}}\int_{-\infty}^{\infty}\left(S_0\exp\left(-\frac{\sigma^2}{2}T+\sigma\sqrt{T}x\right)-Ke^{-rT}\right)_+ e^{-x^2/2}dx$$

$$=\frac{1}{\sqrt{2\pi}}\int_{-\frac{(r-\frac{\sigma^2}{2})T+\ln\frac{S_0}{K}}{\sigma\sqrt{T}}}^{\infty}\left(S_0\exp\left(-\frac{\sigma^2}{2}T+\sigma\sqrt{T}x\right)-Ke^{-rT}\right)e^{-x^2/2}dx$$

$$=\frac{S_0}{\sqrt{2\pi}}\int_{-\frac{(r-\frac{\sigma^2}{2})T+\ln\frac{S_0}{K}}{\sigma\sqrt{T}}}^{\infty}\exp\left(-\frac{\sigma^2}{2}T+\sigma\sqrt{T}x-x^2/2\right)dx$$

$$-\frac{Ke^{-rT}}{\sqrt{2\pi}}\int_{-\frac{(r-\frac{\sigma^2}{2})T+\ln\frac{S_0}{K}}{\sigma\sqrt{T}}}^{\infty}e^{-x^2/2}dx$$

$$=\frac{S_0}{\sqrt{2\pi}}\int_{-\frac{(r+\frac{\sigma^2}{2})T+\ln\frac{S_0}{K}}{\sigma\sqrt{T}}}^{\infty}e^{-y^2/2}dy-\frac{Ke^{-rT}}{\sqrt{2\pi}}\int_{-\infty}^{\frac{(r-\frac{\sigma^2}{2})T+\ln\frac{S_0}{K}}{\sigma\sqrt{T}}}e^{-x^2/2}dx$$

$$=S_0\Phi\left(\frac{\ln(S_0/K)+(r+\sigma^2/2)T}{\sigma\sqrt{T}}\right)-Ke^{-rT}\Phi\left(\frac{\ln(S_0/K)+(r-\sigma^2/2)T}{\sigma\sqrt{T}}\right)$$

as required.

7. $W(t)=W(t)-W(0)$ is a Gaussian random variable with zero mean and variance t. Because this Gaussian distribution is symmetric around 0, the probability that $W(t)>0$ is equal to $1/2$.

8. Answer: $E\left(4W(4)+W(1)\right)=0$ and $Var(4W(4)+W(1))=73$.

9. (i) The process $Y(t)$ is not adapted to the filtration $\{\mathcal{F}_t\}_{t\geq0}$ because at each time t it is not measurable with respect to $\mathcal{F}_t$, it depends on the information about the Wiener paths from t until T.

(ii) The process $Z(t)$ is adapted to the filtration $\{\mathcal{F}_t\}_{t \geq 0}$ because at each time t it is measurable with respect to $\mathcal{F}_t$, it depends on the information about the Wiener paths from 0 until t only.

10. (a) Answer: $Cov(3W(3) + W(2), W(1)) = 4$.

(b) We have $Cov(W(t), W(s)) = E\left(W(t)W(s)\right) - EW(t)EW(s)$. By definition $EW(t) = 0$. Then, using the other parts of the definition and properties of expectations and the Wiener process, we get

$$
\begin{aligned}
Cov(W(t), W(s)) &= E\left(W(t)W(s)\right) \\
&= E\left(W^2(t \wedge s) + W(t \wedge s)(W(t \vee s) - W(t \wedge s))\right) \\
&= Var(W(t \wedge s)) + E\left[W(t \wedge s)(W(t \vee s) - W(t \wedge s))\right] \\
&= t \wedge s + E\left[(W(t \wedge s) - W(0))(W(t \vee s) - W(t \wedge s))\right] \\
&\overset{\text{\scriptsize\it independence of increments}}{=} t \wedge s + E(W(t \wedge s) \\
&\quad - W(0))E(W(t \vee s) - W(t \wedge s)) \\
&= t \wedge s.
\end{aligned}
$$

(c) Answer: $Cov(W(t), W^2(t)) = 0$.

11. We need to check that $\tilde{W}(t)$ satisfies Definition 17.1. Indeed:

1) We have $\tilde{W}(0) = W(s) - W(s) = 0$.

2) For any $0 \leq t_0 < t_1 < \cdots < t_n$, introduce $\tilde{t}_i = t_i + s$. We have $\tilde{W}(t_{i+1}) - \tilde{W}(t_i) = W(t_{i+1} + s) - W(t_i + s) = W(\tilde{t}_{i+1}) - W(\tilde{t}_i)$. Since the increments $W(\tilde{t}_1) - W(\tilde{t}_0), \tilde{W}(\tilde{t}_2) - W(\tilde{t}_1), \ldots, W(\tilde{t}_n) - W(\tilde{t}_{n-1})$ are independent (recall that it is assumed that $W(t)$ is a Wiener process), the increments $\tilde{W}(t_1) - \tilde{W}(t_0), \tilde{W}(t_2) - \tilde{W}(t_1), \ldots, \tilde{W}(t_n) - \tilde{W}(t_{n-1})$ are also independent.

3) We have $\tilde{W}(t) - \tilde{W}(t') = W(t + s) - W(t' + s) \sim \mathcal{N}(0, |t - t'|)$.

4) Since $W(t)$ is a continuous function of t, $\tilde{W}(t) = W(t + s) - W(s)$ is also a continuous function of t.

12. (i) We have (see Problem 2 above):

$$
\varphi(\lambda) = E\left[\exp(\lambda \xi)\right] = \exp\left(\lambda^2 \frac{b^2}{2}\right).
$$

(ii) Recall that $W(t) = W(t) - W(0) \sim \mathcal{N}(0, t)$. We have $\varphi(\lambda) = \exp\left(\lambda^2 \frac{t}{2}\right)$ and

$$
\frac{d^2}{d\lambda^2}\varphi(\lambda) = t \exp\left(\lambda^2 \frac{t}{2}\right) + \lambda^2 t^2 \exp\left(\lambda^2 \frac{t}{2}\right).
$$

Then

$$
\varphi^{(2)}(0) = t.
$$

Further,

$$\frac{d^4}{d\lambda^4}\varphi(\lambda) = t\frac{d^2}{d\lambda^2}\varphi(\lambda) + t^2\frac{d^2}{d\lambda^2}\left[\lambda^2\varphi(\lambda)\right]$$

$$= t\frac{d^2}{d\lambda^2}\varphi(\lambda) + t^2\left[2\varphi(\lambda) + 4\lambda\frac{d}{d\lambda}\varphi(\lambda) + \lambda^2\frac{d^2}{d\lambda^2}\varphi(\lambda)\right].$$

Then

$$E\left[W^4(t)\right] = \frac{d^4}{d\lambda^4}\varphi(0) = t\varphi^{(2)}(0) + 2t^2\varphi(0)$$

$$= t^2 + 2t^2 = 3t^2,$$

as required.

13. Answer: $E\int_0^t W^2(s)dW(s) = 0$ and $E\int_0^t W^2(s)ds = t^2/2$.

14. We have

$$E\left[\int_0^t W_1(s)dW_2(s)\right]^2 = \int_0^t EW_1^2(s)ds = \int_0^t sds = \frac{t^2}{2}.$$

15. Due to the definition of Ito integral, we have

$$\int_0^t W_1(s)dW_2(s) = \underset{h\to 0}{\text{l.i.m.}} \sum_{i=0}^{N-1} W_1(t_i) \times (W_2(t_{i+1}) - W_2(t_i))$$

for a discretization $0 = t_0 < t_1 < \cdots < t_N = t$ with $h = \max(t_{i+1} - t_i)$.

Re-arrange the sum

$$\sum_{i=0}^{N-1} W_1(t_i)(W_2(t_{i+1}) - W_2(t_i))$$

$$= \sum_{i=0}^{N-1}\{-W_2(t_i)(W_1(t_{i+1}) - W_1(t_i))$$

$$+ W_1(t_i)W_2(t_{i+1}) - W_1(t_i)W_2(t_i) + W_2(t_i)W_1(t_{i+1}) - W_2(t_i)W_1(t_i)\}$$

$$= - \sum_{i=0}^{N-1}\{W_2(t_i)(W_1(t_{i+1}) - W_1(t_i))\}$$

$$+ \sum_{i=0}^{N-1}\{W_1(t_i)W_2(t_{i+1}) - W_1(t_i)W_2(t_i)$$

$$+ W_2(t_i)W_1(t_{i+1}) - W_2(t_{i+1})W_1(t_{i+1})\}$$

$$+ \sum_{i=0}^{N-1}[W_2(t_{i+1})W_1(t_{i+1}) - W_2(t_i)W_1(t_i)]$$

$$= -\sum_{i=0}^{N-1} \left\{ W_2(t_i)(W_1(t_{i+1}) - W_1(t_i)) \right\} \qquad (21.5)$$

$$- \sum_{i=0}^{N-1} \left[W_1(t_{i+1}) - W_1(t_i) \right] \left[W_2(t_{i+1}) - W_2(t_i) \right] + W_1(t)W_2(t).$$

Observing that

$$E \sum_{i=0}^{N-1} \left[W_1(t_{i+1}) - W_1(t_i) \right] \left[W_2(t_{i+1}) - W_2(t_i) \right]$$

$$= \sum_{i=0}^{N-1} E \left[W_1(t_{i+1}) - W_1(t_i) \right] \left[W_2(t_{i+1}) - W_2(t_i) \right]$$

$$\overset{independence}{=} \sum_{i=0}^{N-1} E \left[W_1(t_{i+1}) - W_1(t_i) \right] E \left[W_2(t_{i+1}) - W_2(t_i) \right]$$

$$\overset{prop \; of \; W.p.}{=} 0$$

and

$$Var \sum_{i=0}^{N-1} \left[W_1(t_{i+1}) - W_1(t_i) \right] \left[W_2(t_{i+1}) - W_2(t_i) \right]$$

$$\overset{independence}{=} \sum_{i=0}^{N-1} Var \left[W_1(t_{i+1}) - W_1(t_i) \right] \left[W_2(t_{i+1}) - W_2(t_i) \right]$$

$$= \sum_{i=0}^{N-1} E \left[W_1(t_{i+1}) - W_1(t_i) \right]^2 \left[W_2(t_{i+1}) - W_2(t_i) \right]^2$$

$$\overset{independence}{=} \sum_{i=0}^{N-1} E \left[W_1(t_{i+1}) - W_1(t_i) \right]^2 E \left[W_2(t_{i+1}) - W_2(t_i) \right]^2$$

$$\overset{prop \; of \; W.p.}{=} \sum_{i=0}^{N-1} \left[t_{i+1} - t_i \right]^2,$$

which goes to 0 as $h \to 0$, we obtain that

$$\underset{h \to 0}{\text{l.i.m.}} \sum_{i=0}^{N-1} \left[W_1(t_{i+1}) - W_1(t_i) \right] \left[W_2(t_{i+1}) - W_2(t_i) \right] = 0.$$

Combing this fact with (21.5) and the definition of the Ito integral $\int_0^t W_2(s)dW_1(s) = \underset{h \to 0}{\text{l.i.m.}} \sum_{i=0}^{N-1} W_2(t_i) \times (W_1(t_{i+1}) - W_1(t_i))$, we arrive at the required answer.

16. Apply the Ito formula to $sW(s)$ and obtain

$$d(sW(s)) = sdW(s) + W(s)ds,$$

after integrating we arrive at the required formula.

17. Due to the Ito formula, we have

$$d\xi^{(n)} = \frac{\partial}{\partial x}g_n(W(t))dW + \frac{1}{2}\frac{\partial^2}{\partial x^2}g_n(W(t))dt$$

$$= nW^{n-1}(t)dW + \frac{n(n-1)}{2}W^{n-2}(t)dt$$

and thus

$$d\xi^{(n)} = n\xi^{(n-1)}dW + \frac{n(n-1)}{2}\xi^{(n-2)}dt, \quad \xi^{(n)}(0) = 0.$$

Taking expectation from the left- and right-hand sides of the above equation and taking into account that expectation of the Ito integral is zero, we get

$$dE\xi^{(n)} = \frac{n(n-1)}{2}E\xi^{(n-2)}(t)dt.$$

In particular,

$$dE\xi^{(2n)} = \frac{2n(2n-1)}{2}E\xi^{2(n-1)}(t)dt, \quad E\xi^{(2n)}(0) = 0.$$

We have

$$EW^{2n}(t) = E\xi^{(2n)} = \frac{2n(2n-1)}{2}\int_0^t E\xi^{(2n-2)}(s_1)ds_1$$

$$= \frac{2n(2n-1)}{2}\frac{(2n-2)(2n-3)}{2}\int_0^t\int_0^{s_1} E\xi^{(2n-4)}(s_2)ds_2ds_1$$

$$= \frac{2n(2n-1)}{2}\frac{(2n-2)(2n-3)}{2}\cdots\frac{4\times 3}{2}$$

$$\times \int_0^t\cdots\int_0^{s_{n-2}} E\xi^2(s_{n-1})ds_{n-1}\cdots ds_1$$

$$= \frac{(2n)!}{2^n}\int_0^t\cdots\int_0^{s_{n-2}} EW^2(s_{n-1})ds_{n-1}\cdots ds_1$$

$$= \frac{(2n)!}{2^n}\int_0^t\cdots\int_0^{s_{n-2}} s_{n-1}ds_{n-1}\cdots ds_1$$

$$= \frac{(2n)!}{2^n}\frac{t^n}{n!}.$$

18. We have

$$d\ln S = \frac{1}{S}\mu(t)Sdt - \frac{1}{2}\frac{1}{S^2}\sigma^2(t)S^2dt + \frac{1}{S}\sigma(t)SdW(t) =$$

$$= (\mu(t) - \sigma^2(t)/2)dt + \sigma(t)dW(t).$$

Thus,

$$dX = (\mu(t) - \sigma^2(t)/2)dt + \sigma(t)dW(t), \quad X(0) = \ln S_0.$$

19. By the Ito formula, we have

$$dX = -\frac{1}{2}\cos(W(t))dt - \sin(W(t))dW(t).$$

Then

$$dEX = -\frac{1}{2}EX dt, \; EX(0) = 1,$$

and $EX(t) = \exp(-t/2)$ and, in particular, $EX(2) = 1/e$.

20. Analogous to Problems 17–19 and 21.

21. We have

$$dEr = a(b - cEr)dt, \; Er(0) = r_0.$$

Hence

$$Er(t) = \frac{b}{c} + e^{-act}\left(r_0 - \frac{b}{c}\right)$$

and clearly $\lim_{t\to\infty} Er(t) \to \frac{b}{c}$.

By the Ito formula, we have

$$dr^2 = 2a(b - cr)rdt + \sigma^2 rdt + 2r\sigma\sqrt{r}dW(t), \; r^2(0) = r_0^2.$$

Hence

$$dEr^2 = \left(2ab + \sigma^2\right)Er - 2acEr^2 dt, \; Er^2(0) = r_0^2.$$

Then $\rho(t) = Er^2$ satisfies

$$\dot\rho = \frac{2ab^2 + \sigma^2 b}{c} + e^{-act}\left(r_0 - \frac{b}{c}\right)\left(2ab + \sigma^2\right) - 2ac\rho, \; \rho(0) = r_0^2.$$

Hence

$$Er^2 = \frac{b\sigma^2}{2ac^2} + \frac{b^2}{c^2} + \left(r_0 - \frac{b}{c}\right)\left(\frac{2b}{c} + \frac{\sigma^2}{ac}\right)e^{-act}$$

$$+ \left(r_0^2 + \frac{b\sigma^2}{2ac^2} + \frac{b^2}{c^2} - r_0\frac{2b}{c} - r_0\frac{\sigma^2}{ac}\right)e^{-2act}.$$

Thus,

$$Var(r(t)) = \frac{b\sigma^2}{2ac^2} + \left(r_0 - \frac{b}{c}\right)\frac{\sigma^2}{ac}e^{-act} + \frac{\sigma^2}{ac}\left(\frac{b}{2c} - r_0\right)e^{-2act}.$$

22. We first transform the problem (18.42)–(18.43) to the Cauchy problem for heat equation. To this end, introduce new independent variables s and y and the new function $u(s, y)$ so that

$$s = (T - t)\sigma^2, \quad y = \ln x + \frac{r - \sigma^2/2}{\sigma^2} s$$

and

$$v(t, x) = e^{rs(t)/\sigma^2} u(s(t), y(t, x)).$$

Applying elementary Calculus, we can show (please do this exercise) that $u(s, y)$ is the solution of the following Cauchy problem for heat equation:

$$\frac{\partial u}{\partial s} = \frac{1}{2} \frac{\partial^2 u}{\partial y^2}, \ t > 0, \ y \in \mathbf{R}, \tag{21.6}$$

$$u(0, y) = e^{rs/\sigma^2} f(e^y).$$

The problem (21.6) can be solved by any standard method, e.g., via knowing its fundamental solution or via Fourier transform (see, e.g., Fridman, 1964; Evans, 2010). Let us use the former. The fundamental solution is

$$\rho(s, y) = \frac{1}{\sqrt{2\pi s}} e^{-y^2/2s}$$

and then the solution of (21.6) can be written as

$$u(s, y) = \frac{1}{\sqrt{2\pi s}} \int_{-\infty}^{\infty} u(0, z) e^{-z^2/2s} dz = \frac{e^{rs/\sigma^2}}{\sqrt{2\pi s}} \int_{-\infty}^{\infty} f(e^z) e^{-z^2/2s} dz. \tag{21.7}$$

Substituting the given payoff $f(x) = (x - K)_+$ in (21.7), performing manipulations with the integral (cf. Problem 6 above) and finally making the reverse change of variables, one arrives (please do these calculations) at the required (18.44).

23. (a) Introduce

$$dX = dW(s), \quad X(t) = x.$$

Then the solution $u(t, x)$ has the following probabilistic representation

$$u(t, x) = E\left[X_{t,x}(1) \exp(-c(1 - t))\right], \ t \in [0, 1].$$

(b) We have

$$u(t, x) = E\left[X_{t,x}(1) \exp(-c(1 - t))\right] = \exp(-c(1 - t)) E(x + W(1) - W(t))$$
$$= x \exp(-c(1 - t)).$$

24. (a) See Section 18.3.

(b) We have

$$
\mathcal{V} = \frac{\partial}{\partial \sigma} C_0 = S_0 \frac{\partial}{\partial \sigma} \Phi \left(\frac{\ln \frac{S_0}{K} + (r + \frac{\sigma^2}{2})T}{\sigma \sqrt{T}} \right)
$$

$$
- K e^{-rT} \frac{\partial}{\partial \sigma} \Phi \left(\frac{\ln \frac{S_0}{K} + (r - \frac{\sigma^2}{2})T}{\sigma \sqrt{T}} \right)
$$

$$
= S_0 \left(-\frac{\ln \frac{S_0}{K} + rT}{\sigma^2 \sqrt{T}} + \frac{\sqrt{T}}{2} \right) \Phi' \left(\frac{\ln \frac{S_0}{K} + (r + \frac{\sigma^2}{2})T}{\sigma \sqrt{T}} \right)
$$

$$
- K e^{-rT} \left(-\frac{\ln \frac{S_0}{K} + rT}{\sigma^2 \sqrt{T}} - \frac{\sqrt{T}}{2} \right) \Phi' \left(\frac{\ln \frac{S_0}{K} + (r - \frac{\sigma^2}{2})T}{\sigma \sqrt{T}} \right)
$$

$$
= \frac{S_0 \sqrt{T}}{\sqrt{2\pi}} \exp \left[-\frac{1}{2} \left(\frac{\ln \frac{S_0}{K} + (r + \frac{\sigma^2}{2})T}{\sigma \sqrt{T}} \right)^2 \right].
$$

(c) Because in the Black–Scholes world prices are positive and the exponent is always positive, $\mathcal{V} > 0$.

(d) The positiveness of the vega means that, assuming that all the other parameters are kept the same, plain vanilla calls have a higher value when volatility of the underlying is larger. Economically, we can understand this as follows. The higher the volatility, the more risky the underlying is and it is more risky to write an option on this underlying. Consequently, the writer charges a higher premium for options on an underlier with larger volatility.

25. (a) Answer:

$$
\Delta = -\Phi \left(\frac{\ln(K/S_0) - (r + \sigma^2/2)T}{\sigma \sqrt{T}} \right).
$$

(b) Obviously, Δ is always negative.

(c) The negativeness of the delta means that, assuming that all the other parameters are kept the same, plain vanilla puts have a lower value when price of the underlying stock is larger. Economically, we can understand this as follows. The larger price of the underlier, the less attractive the put is (recall that a put's payoff is $(K - s)_+$). Consequently, the premium of a put on an underlier with a larger price is smaller. The other implication is that in the Black–Scholes world hedging of a put always requires borrowing of the underlier.

26. Hint. Use the put–call parity.

27. Answer: $\Gamma \to \infty$ as $T - t \to 0$ for an ATM call or put.

28. Analogous to Example 18.11.

Bibliography

Andersen, L. and Piterbarg, V. (2010). *Interest Rate Modeling. Volumes 1, 2 and 3* (Atlantic Financial Press, London, UK).

Applebaum, D. (2009). *Levy Processes and Stochastic Calculus* (Cambridge University Press, Cambridge, UK).

Arnold, V. I. (1983). *Geometrical Methods in the Theory of Ordinary Differential Equations* (Springer-Verlag, Berlin Heidelberg, Germany).

Avellaneda, M. and Laurence, P. (2000). *Quantitative Modeling of Derivative Securities: From Theory to Practice* (Chapman & Hall/CRC Press, Boca Raton, FL, USA).

Bachelier, L. (1900). Théorie de la Spéculation, *Annales Scientifique de l'Ecole Normale Supérieure* 3e *série* **17**, pp. 21–86.

Bank for International Settlements (2012). Statistical release: OTC derivatives statistics at end-December 2011, [Online] http://www.bis.org/publ/otc_ny1205.pdf [Accessed 26 April 2013].

Barucci, E. (2003). *Financial Markets Theory* (Springer-Verlag, New York, NY, USA).

Baxter, M. and Rennie, A. (1996). *Financial Calculus* (Cambridge University Press, Cambridge, UK).

Bernstein, S. N. (1934). Principes de la theorie des equationnes differentielles stochastiques, *Travaux de l'Institut Physicow Mathematique de Stekloff* **5**, pp. 95–124.

Bielecki, T. R. and Rutkowski, M. (2010). *Credit Risk: Modeling, Valuation and Hedging* (Springer-Verlag, Berlin Heidelberg, Germany).

Billingsley, P. (1968). *Convergence of Probability Measures* (Wiley, New York, NY, USA).

Bingham, N. H. and Kiesel, R. (2004). *Risk-Neutral Valuation. Pricing and hedging of financial derivatives* (Springer-Verlag, London, UK).

Björk, T. (2004). *Arbitrage Theory in Continuous Time* (Oxford University Press, New York, NY, USA).

Björk, T., Kabanov, Y. and Runggaldier, W. (1997). Bond market structure in the presence of marked point processes, *Mathematical Finance* **7**, pp. 211–239.

Black, F. and Scholes, M. (1973). The pricing of options and corporate liabilities, *Journal of Political Economy* **81**, pp. 637–654.

Boyle, M. (2006). *The Art of Bookmaking: How to Compile Odds for Any Sporting Event* (High Stakes Publishing, London, UK).

Brigo, D. and Mercurio, F. (2006). *Interest Rate Models – Theory and Practice: With Smile, Inflation and Credit* (Springer-Verlag, Berlin Heidelberg, Germany).

Brown, R. (1828). A brief account of microscopical observations made in the months of June, July and August 1827, on the particles contained in the pollen of plants; and on the general existence of active molecules in organic and inorganic bodies, *Philosophical Magazine Series 2* **4**, pp. 161–173.

Capinski, M. and Zastawniak, T. (2003). *Mathematics for Finance: An Introduction to Financial Engineering* (Springer-Verlag, London, UK).

Carr, P., Geman, H., Madan, D. and Yor, M. (2002). The fine structure of asset returns: An empirical investigation, *Journal of Business* **75**, pp. 305–332.

Central Intelligence Agency (2013). The World Factbook, [Online] https://www.cia.gov/library/publications/the-world-factbook/geos/xx.html [Accessed 26 April 2013].

Cesari, G., Aquilina, J., Charpillon, N., Filipovic, Z., Lee, G. and Manda, I. (2010). *Modelling, Pricing, and Hedging Counterparty Credit Exposure: A Technical Guide* (Springer-Verlag, Berlin Heidelberg, Germany).

Cheney, E. W. and Kincaid, D. R. (2012). *Numerical Mathematics and Computing* (Brooks/Cole-Thomson Learning, Belmont, CA, USA).

Choudhry, M. (2012). *The Principles of Banking* (Wiley, New York, NY, USA).

Cont, R. and Tankov, P. (2003). *Financial Modelling with Jump Processes* (Chapman & Hall/CRC Press, Boca Raton, FL, USA).

Courtault, J.-M., Kabanov, Y., Bru, B., Crépel, P., Lebon, I. and Marchand, A. L. (2000). Louis Bachelier: On the centenary of Théorie de la Spéculation, *Mathematical Finance* **10**, pp. 341–353.

Cox, J. C., Ross, S. and Rubinstein, M. (1979). Option pricing: a simplified approach, *Journal of Financial Economics* **7**, pp. 229–263.

Cox, J. C. and Rubinstein, M. (1985). *Options Markets* (Prentice Hall, Upper Saddle River, NJ, USA).

Davis, M. and Etheridge, A. (2006). *Louis Bachelier's Theory of Speculation. The Origins of Modern Finance* (Princeton University Press, Princeton, NJ, USA).

Delbaen, F. (2003). Risk measures or measures that describe risk? Tech. rep., ETH, [Online] http://www.math.ethz.ch/~delbaen/ftp/preprints/Riskexecutives.pdf [Accessed 14 March 2013].

Derman, E. (2004). *My Life as a Quant: Reflections on Physics and Finance* (Wiley, New York, NY, USA).

Dokuchaev, N. (2007). *Mathematical Finance: Core Theory, Problems and Statistical Algorithms* (Taylor & Francis, Abingdon, UK).

Duffy, D. J. (2006). *Introduction to C++ for Financial Engineers* (Wiley, Chichester, UK).

Einstein, A. (1905). Über die von der molekularkinetischen Theorie der wärme gefordete Bewegung von in ruhenden Flüssigkeiten suspendierten Teilchen, *Annalen der Physik* **17**, pp. 132–148.

Elton, E. J., Gruber, M. J., Brown, S. J. and Goetzmann, W. N. (2009). *Modern Portfolio Theory and Investment Analysis* (Wiley, Hoboken, NJ, USA).

Etheridge, A. (2002). *A Course in Financial Calculus* (Cambridge University Press, Cambridge, UK).

Evans, L. C. (2010). *Partial Differential Equations* (American Mathematical Society, Providence, RI, USA).

Fengler, M. R. (2009). *Semiparametric Modeling of Implied Volatility* (Springer-Verlag, Berlin Heidelberg, Germany).

Filipovic, D. (2009). *Term-Structure Models. A Graduate Course* (Springer-Verlag, Berlin Heidelberg, Germany).

Föllmer, H. and Schied, A. (2004). *Stochastic Finance: An Introduction in Discrete Time* (Walter de Gruyter, Berlin, Germany).

Ford, B. J. (1992). Brownian movement in clarkia pollen: a reprise of the first observations, *Microscope* **40**, pp. 235–241.

Fridman, A. (1964). *Partial Differential Equations of Parabolic Type* (Prentice-Hall International, Englewood Cliffs, NJ, USA).

Fristedt, B., Jain, N. and Krylov, N. (2007). *Filtering and Prediction: a primer* (American Mathematical Society, Providence, RI, USA).

Glasserman, P. (2003). *Monte Carlo Methods in Financial Engineering* (Springer-Verlag, New York, NY, USA).

Harrison, J. M. and Pliska, S. R. (1981). Martingales and stochastic integrals in the theory of continuous trading, *Stochastic Processes and their Applications* **11**, pp. 215–260.

Hasminskii, R. Z. (1980). *Stochastic Stability of Differential Equations* (Sijthoff & Noordhoff, Alphen aan den Rijn, Netherlands).

Higham, D. J. (2004). *An Introduction to Financial Option Valuation* (Cambridge University Press, Cambridge, UK).

Holden, H., Oksendal, B., Uboe, J. and Zhang, T. (2010). *Stochastic Partial Differential Equations. A Modelling, White Noise Functional Approach* (Springer-Verlag, Berlin Heidelberg, Germany).

Holton, G. A. (2004). Defining risk, *Financial Analysts* **60**, pp. 19–25.

Hull, J. C. (2003). *Options, Futures and Other Derivatives* (Prentice Hall, Upper Saddle River, NJ, USA).

Hull, J. C. (2006). *Options, Futures and Other Derivatives: Student Solutions Manual* (Prentice Hall, Upper Saddle River, NJ, USA).

Ito, K. (1942). Differential equations determining a markov process (in Japanese), *Zenkoku Shijo Danwakai* **1077**, pp. 1352–1400.

Ito, K. (1944). Stochastic integral, *Proceedings of the Imperial Academy, Tokyo* **20**, pp. 519–524.

Ito, K. and McKean, H. R. (1965). *Diffusion Processes and Their Sample Paths* (Springer-Verlag, Berlin Heidelberg, Germany).

Jarrow, R. and Protter, P. (2004). A short history of stochastic integration and mathematical finance: the early years, 1880–1970, *IMS Lecture Notes* **45**, pp. 75–91.

Jewson, S. and Brix, A. (2005). *Weather Derivative Valuation: the Meteorological, Statistical, Financial and Mathematical Foundations* (Cambridge University Press, Cambridge, UK).

Jorion, P. (2006). *Value at Risk* (McGraw-Hill Profesional, New York, NY, USA).

Joshi, D. J. (2008). *C++ Design Patterns and Derivatives Pricing* (Cambridge University Press, Cambridge, UK).

Karatzas, I. and Shreve, S. E. (1991). *Brownian Motion and Stochastic Calculus* (Springer-Verlag, Berlin Heidelberg, Germany).

Karatzas, I. and Shreve, S. E. (1998). *Methods of Mathematical Finance* (Springer-Verlag, New York, NY, USA).

Kindleberger, C. P. (2000). *Manias, Panics, and Crashes: A History of Financial Crises* (Wiley, New York, NY, USA).

Klebaner, F. C. (2005). *Introduction to Stochastic Calculus with Applications* (Imperial College Press, London, UK).

Kolb, R. W. (2003). *Futures, Options, and Swaps* (Blackwell, Oxford, UK).

Kolmogorov, A. N. (1931). Über die analytischen methoden in der Wahrscheinlichkeitsrechnugen, *Mathematische Annalen* **104**, pp. 415–458.

Kolmogorov, A. N. (1933). *Grundbegriffe der Wahrscheinlichkeitsrechnung* (Springer-Verlag, Berlin Heidelberg, Germany).

Krylov, N. V. (2002). *Introduction to the Theory of Random Processes* (American Mathematical Society, Providence, RI, USA).

Kwok, Y.-K. (2008). *Mathematical Models of Financial Derivatives* (Springer-Verlag, Berlin Heidelberg, Germany).

Lamberton, D. and Lapeyre, B. (2007). *Introduction to Stochastic Calculus Applied to Finance, 2nd edition* (Chapman & Hall/CRC Press, Boca Raton, FL, USA).

Levy, H. (2006). *Stochastic Dominance - Investment Decision Making Under Uncertainty* (Springer Science+Business Media, New York, NY, USA).

Lipton, A. and Rennie, A. (eds.) (2008). *The Oxford Handbook of Credit Derivatives* (Oxford University Press, Oxford, UK).

Lo, A. W. and MacKinlay, C. (2002). *A Non-Random Walk Down Wall Street* (Princeton University Press, Princeton, NJ, USA).

Mao, X. (2007). *Stochastic Differential Equations and Applications* (Horwood Publishing, Chichester, UK).

Melnikov, A. (2004). *Risk Analysis in Finance and Insurance* (Chapman & Hall/CRC Press, Boca Raton, FL, USA).

Merton, R. (1973). Theory of rational option pricing, *Bell Journal of Economics and Management Science* **4**, pp. 141–183.

Meyer, P. L. (1970). *Introductory Probability and Statistical Applications* (Addison Wesley, Boston, MA, USA).

Milstein, G. N. and Schoenmakers, J. G. M. (2002). Monte Carlo construction of hedging strategies against multi-asset european claims, *Stochastics and Stochastics Reports* **73**, pp. 125–157.

Milstein, G. N. and Tretyakov, M. V. (2004). *Stochastic Numerics for Mathematical Physics* (Springer-Verlag, Berlin Heidelberg, Germany).

Montroll, E. W. and Shloringer, M. F. (1984). On the wonderful world of random walks, in J. L. Lebowitz and E. W. Montroll (eds.), *Nonequilibrium Phenomena II: from Stochastics to Hydrodynamics* (North-Holland Publishing, Amsterdam, The Netherlands), pp. 1–121.

Musiela, M. and Rutkowski, M. (2005). *Martingale Methods in Financial Modelling* (Springer-Verlag, Berlin Heidelberg, Germany).

Paul, W. and Baschnagel, J. (1999). *Stochastic Processes From Physics to Finance* (Springer-Verlag, Berlin Heidelberg, Germany).

Pliska, S. R. (1997). *Introduction to Mathematical Finance: Discrete Time Models* (Blackwell Publishers, Malden, MA, USA).

Rogers, L. C. G. and Williams, D. (2000). *Diffusions, Markov Processes and Martingales. Volume 1, Foundations* (Cambridge University Press, Cambridge, UK).

Ruppert, D. (2011). *Statistics and Data Analysis for Financial Engineering* (Springer Science+Business Media, New York, NY, USA).

Samuelson, P. (1965). Rational theory of warrant pricing, *Industrial Management Review* **6**, pp. 13–39.

Schlogl, E. (2013). *Quantitative Finance: An Object-Oriented Approach in C++* (Chapman & Hall/CRC Press, Boca Raton, FL, USA).

Schmid, B. (2004). *Credit Risk Pricing Models: Theory and Practice* (Springer-Verlag, Berlin Heidelberg, Germany).

Schoenmakers, J. (2005). *Robust Libor Modelling and Pricing of Derivative Products* (Chapman & Hall/CRC Press, Boca Raton, FL, USA).

Shiryaev, A. N. (1996). *Probability* (Springer-Verlag, New York, NY, USA).

Shiryaev, A. N. (1999). *Essentials of Stochastic Finance: Facts, Models, Theory* (World Scientific Publishing, Singapore).

Shreve, S. E. (2003). *Stochastic Calculus for Finance I: the Binomial Asset Pricing Model* (Springer Science+Business Media, New York, NY, USA).

Shreve, S. E. (2004). *Stochastic Calculus for Finance II: Continuous-Time Models* (Springer Science+Business Media, New York, NY, USA).

Stefanica, D. (2011). *A Primer for the Mathematics of Financial Engineering* (Financial Engineering Press, New York, NY, USA).

Thomson, A. (2012). UK treasury minister: Government accepts recommendations of Wheatley Libor review in full, *Dow Jones Newswires* 17 October 2012.

Wackerly, D. D., Mendenhall, W. and Scheaffer, R. L. (2007). *Mathematical Statistics with Applications* (Duxbury Press, Belmont, CA, USA).

Wentzell, A. D. (1981). *A Course in the Theory of Stochastic Processes* (McGraw-Hill International, New York, NY, USA).

Wiener, N. (1923). Differential space, *Journal of Mathematical Physics* **2**, pp. 131–174.

Williams, D. (2001). *Weighing the Odds* (Cambridge University Press, Cambridge, UK).

Wilmott, P. (2008). Science in Finance IX: In Defence of Black, Scholes and Merton, [Online] http://www.wilmott.com/blogs/paul/index.cfm/2008/4/29/ Science-in-Finance-IX-In-defence-of-Black-Scholes-and-Merton   [Accessed 14 March 2013].

Zhou, X. (2008). *A Practical Guide To Quantitative Finance Interviews* (Lulu Press, Raleigh, NC, USA).

Index

Admissible strategy, 103
American options, 125
Arbitrage, 20
  free market, 20
  price, 20
Asian options, 107
Asset, 14
At-the-money (ATM), 46
Attainable claim, 107

Base point, 39
Bears, 51
Bermudian option, 136
Binomial representation theorem, 98, 111
Binomial tree, 68
Black–Scholes equation, 201, 204
Black–Scholes formula
  continuous, 171, 174
  discrete, 113, 173
Bond, 25
  coupon, 34
  default-free zero-coupon discount, 25
  face value, 25
  principal, 25
Borel sigma-algebra, 161
British Banker's Association (BBA), 35
Brownian motion, 179, 182
Bull spread, 60
Bulls, 50

Butterfly spread, 60

Calibration, 175
Central limit theorem, 170
Chooser option, 61
Clearing house, 23
Close-of-trading price, 155
Complete market, 120
Concave function, 148
Consensus market prices, 211
Continuation value, 133
Convex function, 137

Delta, 112, 206
Delta hedging, 112, 207
Density, 164
Derivatives, 15
Discount curve, 26
Discounted price process, 96, 102
Distribution function, 80, 163, 164
  joint, 165
Drift, 156

Efficient market hypothesis, 154
Elementary probabilities, 77
Equivalent martingale measure, 102
Equivalent probability measures, 101
European option, 43
Exotic options, 107, 145
Expectation, 80, 165
  conditional, 84, 91

Face value, 25